WHERE, OH WHERE DID THE STAR OF BETHLEHEM GO?

Philip Rastocny

Philip Rastocny

Where, Oh Where did the Star of Bethlehem Go?

By Philip Rastocny

Amazon edition

Publisher: Grasslands Publishing House, 8526 Central Avenue, Brooksville, Florida 34613

ISBN: 978-0-9854081-0-7

First edition — December, 2014
Second edition — April, 2015
Third edition — November, 2015
Fourth edition – October, 2016

Discover other titles by Philip Rastocny at http://www.Amazon.com

Cover graphics and design by Philip Rastocny

Editing by Althea Rose

All photographs by Althea Rose or Philip Rastocny, or public domain *except*

Photographs of SN1987a before/after exploding by David Malin, Australian Astronomical Observatory, used with permission

The astronomy software used in developing images was *The Sky*© *VI* created by Software Bisque, 862 Brickyard Circle, Golden, CO 80403-8058 USA.

Retrograde information provided by http://www.astropro.com

Where, Oh Where did the Star of Bethlehem Go?

For my amazing soul mate

Althea Rose

Philip Rastocny

Table of Contents

Table of Contents
Forward
Introduction
When Are We?
Look, Up in the Sky…
What Did Ancient Astronomer Do All Night?
What Was So Important?
   Eliminating Unlikely Suspects
   The Usual Suspects
   What is Retrograde Motion?
   Case Review
There's No Place like Home…
In the Beginning…
   A Star's Relative Brightness
   Summary
One if by Land, Two if by Sea…
Water and Land Trade Routes of 4 BCE
Twinkle, Twinkle, Little Star…
   The Shepherd's Field
Exit…Stage Right…
What Did the Original Words Say?
What Would You Carry?
Reviewing the Facts
   What Does This All Mean?
Where Is This Star Today?
About the Author

Forward

This book is a combined work of science, conjecture, fancy, and intuition. You see, it has to be this way since nothing is more difficult to prove than ancient history. I bet you cannot recall what you ate last Wednesday night much less who ran for Vice President in the 1996 Presidential election. So much time has passed and so many historical documents lost that few remaining reliable texts survive to support any real facts or conclusions. Speculation, even reliable innuendo, is at best a pipe dream…and that is exactly what this book is…more pipe dreams.

However, much like solving a good mystery, the same techniques can be applied when trying to figure out what in the world (or out of this world) happened to the so called "Star of Bethlehem." Many people get wrapped up in proving this and disproving that because of these issues and those facts, but the truth is this: first, nobody really knows and second, does it really matter?

I have a confession to make at the outset so that you can bear with me (you will see what I mean in some parts of this book when I get long winded). I am a member of P.A., better known as *Pedantics Anonymous*, a little-known self-help group of individuals, whose purpose is much like that of the well-known self-help group *Alcoholics Anonymous* (A.A.) P.A. helps those individuals who love to go on and on about the details of things, often boring but nonetheless usually true, much like helps those recovering from alcoholism.

PE-DAN-TIC adjective: Narrowly, stodgily, and often ostentatiously learned.

Pedantics are those folks, typically academics, who sometimes testify as so called experts in legal proceedings because they appear to know everything about one particular subject. I, as an upstanding member of P.A., can testify that in many meetings one person talks about the meaning of life as seen through the eyes of a tsetse fly (fascinating if you get my drift).

So I will try to keep the Pedantic stuff down to a minimum while keeping you entertained. Who knows, maybe I'll see you in my next P.A. meeting?

BTW, to the best of my knowledge there is no such thing as Pedantics Anonymous; it was just a joke. I need to clarify this so that those who are taking what I say seriously can understand a little about my somewhat dry humor. My tendency to ramble, however, is accurate and I am trying to get that part of me under control (and seriously, this is no joke).

Introduction

When I was a young boy in the dark rural suburbs of a small town in Wisconsin, I used to lay down in the fresh cut grass in our backyard on warm summer nights gazing up at the black sky with total fascination and wonder. The stars all seemed so far away and the sounds of the crickets intensified my mystification with these unimaginable distances. Everything I saw, all of those tiny twinkling pinpoints of light, were beyond the ability of my young mind to conceive and put into perspective with how far I needed to bicycle to school. They all seemed beyond my reach in many ways but yet there they were, unmoving, brilliant, and so captivating.

I started to study astronomy shortly after looking through a neighbor's telescope and seeing the rings of Saturn with my own eyes. Seeing pictures in books and experiencing a spectacular celestial object firsthand was much like the difference between reading about how to drive a car and then driving a car; they were two completely different experiences. I was hooked but it took another 10 years for me to get my first telescope.

I went to a conservative parochial school from kindergarten through high school and was taught much about the Bible and its wisdom. Spending an hour a day under the careful instruction of trained professionals will give you insights that can change your life for the better. But when it came to using the contents of the Bible as a reference of scientific fact, well let's say that I had many unanswered questions that my best teachers could not satisfy. There seemed to be a paradox between a literal interpretation of these Holy Words and what I saw with my own eyes (much like the difference between looking at pictures of Saturn and seeing it through a telescope). While this did not alter my faith, I chose to seek answers to my burning questions outside of the Church.

As a person fascinated with the night sky, one of these burning questions was: *What did the Magi see in the sky that caused them to travel from their home in the East to a tiny town along the Mediterranean coast?* It is this burning question that is the topic of this book.

Other questions arose in my journey to understand the answer to this fundamental unanswered burning question that I also hope to disclose. These associated questions are:

- Who were these ancient astronomers called the Magi?
- Why would they undertake such a long journey?
- What did these ancient astronomers know that eludes common sense today?

Where, Oh Where did the Star of Bethlehem Go?

- Where did the Magi call home?
- What route did they travel to get to the Mediterranean Sea?
- Where did they stand to see what they came to see?

Many people have tried to explain what they believe to be this single ever-puzzling event. Some believe it was a comet, others a planetary alignment or conjunction, others a shooting star, and still others a divinely created one-time event. The truth is that no one can prove what this astronomical event actually was and because of this speculation bounds.

However, I am a bit different from those people who leverage their beliefs to fit science; I know that nature is a reflection of the hand of its Creator and that there is much we can tell about the Creator by the bread crumbs left in what we call science and mathematics. It is the rules of physics and patterns in nature that are these bread crumbs and all else is pure speculation.

With this perspective, much like the forensics used to solve murder mysteries, that I will approach the answer to my burning question. With my passion in astronomy and my logical mind, I believe I have found the answer to this question. Regardless of your beliefs, I invite you to come along with me on this fascinating journey into what we know as truth and what we read as supposition. Set your biases aside for a time and consider what may have happened as postulated by this new and interesting approach. Know that perhaps what I propose is as much fiction as it is fact, but then from my best research no one has taken this question to this extreme multi-level of analysis as I.

Give it a chance and think about it. I am hopeful that what you read may reveal to you a new light in your heart and one that could strengthen your beliefs as a result. For me, it's nice to know that what I believe can be confirmed in another way. Putting it another way, it reassures me that the core of my beliefs is implied to be true from another perspective.

From this brief introduction, there are many associated questions that must also be answered along the way. Let's start now by understanding where this all began.

> *When I'm drivin' in my car*
> *And that man comes on the radio*
> *He's telling' me more and more*
> *About some useless information*
> *Supposed to fire my imagination...*

Lyric excerpt from the song *Satisfaction* by The Rolling Stones

Take your time, do good work, and have fun.

Philip Rastocny

When Are We?

In 1999, there was an issue with computers where someone forgot to account for longevity (building something that was going to last more than a few years). For computers, it was their built-in ability to NOT to account for a year whose date began with something other than the digit "1" (e.g., 1984, 1990, 1999, etc.). This is where my search gained traction (What the...? Patience...all will become clear...this story is a process, not an event).

Y2K

The most brilliant minds of that time – those who spent years developing computers and programs – literally overlooked the longevity of their equipment. They were so busy making sure that everything worked they forgot to consider the calendar year changing the first digit from "1" to "2." The result of this oversight was that anything that relied on knowing the correct day would literally become confused after December 31, 1999 and most likely stop working. Oops! Sorry about that!

In what was known as the "year 2000 bug" (affectionately called the Y2K bug) a massive barrage of programmers and technicians resolved just about every issue on every computer and in every program anyone could predict. As a result of this monumental last-gasp effort, a few traffic lights in downtown Cincinnati may have stopped working for a few days (they eventually got on top of those too) but the world did not end as some idiots predicted. Doomsday was averted. Whoops! Sorry about that.

But many people also began to make some other relevant queries, one of which I found most intriguing: *What Year Is it?* People began to question if the current year was the correct year. For example, a parishioner in the Catholic Church wrote Zenit *Magazine* (http://www.zenit.org) asking the question "In What Year was Christ Actually Born?" The response, reproduced by permission, is shown below although for some unknown reason the posting on their website is no longer available (you can still find this article listed by searching for it as "in what year was Christ actually born" "rome, dec 14").

From http://www.catholiceducation.org/articles/facts/fm0004.html

Question: In What Year was Christ Actually Born?

ROME, DEC 14 (ZENIT) - While Jubilee celebrations started with the opening of the Holy Door on December 24, 1999 and ended with its closing on January 6, 2000, it is less clear whether this is really the 2000th anniversary of Christ's birth.

When Dionysius Exiguus computed the date of Christ's birth in the Middle Ages, he named the year of the Nativity 1 A.D., and stated that Jesus' birth date was December 25 of that year. The year immediately before this was the year 1 B.C. Whether from mathematical ignorance or design, he did not include a year zero.

This complicates the calculation of the dates of the Jubilee. Christmas of the year 2 A.D. was the 1st anniversary of Christ's birth, according to Dionysius' calculations; similarly, the second anniversary of that birth fell in the year 3 A.D. Taking this forward a few centuries, we find that the 2000th anniversary of Christ's birth should fall on December 25, 2001.

To complicate matters further, it seems that Dionysius' made an error in his calculations. Herod the Great, who the Bible says was alive at the time of Christ's birth, died in the year 4 B.C., based on the reports of Josephus. According to the Gospel of Matthew, when Herod was unable to trick the astrologers into leading him to the Child, he ordered the slaughter of all the male babies in Bethlehem. Since Herod's command (which is not attested outside the Gospels, but is consistent with his historical character) was to kill all babies under age 2, this event occurred no more than 2 years after Christ's birth. If we assume that this happened near the end of Herod's life (which seems likely), this puts Christ's birth in the year 5 or 6 B.C.

In that case, the 2000th anniversary of Christ's birth has already gone by, having been in 1993 or 1994. Naturally at the distance of years, it is practically impossible to say with certainty what year Christ was actually born, though sometime between 7 B.C. and 1 B.C. seems all but certain.

If this is the case, why is the Church celebrating the year 2000 with such solemnity? The answer is simple: because the world is celebrating this date. In his Apostolic Letter "Tertio Millennio Adveniente," the Holy Father wrote, "In view of this, the two thousand years which have passed since the Birth of Christ (prescinding from the question of its precise chronology) represent an extraordinarily great Jubilee, not only for

Christians but indirectly for the whole of humanity, given the prominent role played by Christianity during these two millennia. It is significant that the calculation of the passing years begins almost everywhere with the year of Christ's coming into the world, which is thus the centre of the calendar most widely used today. Is this not another sign of the unparalleled effect of the Birth of Jesus of Nazareth on the history of mankind?"

For the Pope, the year 2000 is a sign of the centrality of Christianity in our society, hence it is cause for celebration and Jubilee. ZE99121521

ACKNOWLEDGEMENT

ZENIT is an International News Agency based in Rome whose mission is to provide objective and professional coverage of events, documents and issues emanating from or concerning the Catholic Church for a worldwide audience, especially the media.

Copyright © 1999 ZENIT

From this article, what is important to note is the excerpt about the actual year being "**…sometime between 7 B.C. and 1 B.C. seems all but certain…**" However, the understandably limited response in this ZENIT excerpt could not embrace the full reasons behind their response to this seemingly simple question. To completely understand why this is not a simple question to answer, you must understand a bit about how calendars are created so the pedantic part of me gets to come out and play now for a little while. Stay tuned because if you bear with me all of this will eventually make sense and it will explain what happened to the Star of Bethlehem (remember, that's why you wanted to buy this book in the first place, right?).

The calendar we currently use is called the *Gregorian* calendar (also known as the *Christian, Western,* or *Civil* calendar). It was commissioned in 1582 by Pope Gregory XIII (hence the name) to replace the calendar in use at the time called the *Julian Calendar* named after Julius Caesar of ancient Rome. The Julian calendar had been in use since 45 BCE and its significance is that it was the first calendar to use 12 months and a "leap" day in February. You see it takes the earth exactly 365.2426 days to make one orbit around the sun (roughly 365 and ¼ days) so a day needed to be added every four years to keep everything in sync. More on this later but for now know that not using a calendar with this adjustment throws the dates for planting crops off and with enough time passing seasonal changes do not synchronize to the calendar date (spring becomes fall).

The significance of the change by the Gregorian calendar rooted in the desire for the Catholic Church to use the date of the Birth of Jesus Christ as the starting year

(year 0001). It's not that the Julian calendar did not work; it's just another fine example of useless politics. A man named Alosyius Lilius, a respected doctor of the time, proposed this idea of a new calendar to the Pope. The Pope apparently liked the idea of his name being assigned to the calendar for all time (another fine example of how the ego gets involved in some decisions). As an aside, if the Pope wanted to eliminate his personal vanity he could have called it the *Jesus* rather than the *Gregorian calendar.*

Pope Gregory knew what was riding on the accuracy of a new calendar so he assigned the task for figuring out precisely how many years had elapsed since the birth of Jesus to the most respected mathematician in the Catholic Church at that time, a monk named Dionysius Exiguus. After all, Pope Gregory would have been mercilessly criticized for making any form of error so he wisely recruited someone else to take the blame in the event such an error was made.

The problem arose when trying to map the recorded date of Easter Sunday (based on the name of the Pope presiding at the **Paschal Sunday** service) into the new Gregorian calendar. What on the surface seems like a no-brainer was a bit more complicated because of the way the records were created. The original table (see sample below) used the date on which the full moon fell as a reference point and then assigned the Paschal Sunday as the first Sunday after that event, a logical date independent of politics or vanity. Compounding this unusual accounting method was the fact that Paschal Sunday does not always fall on the same calendar day.

Year	Indiction	Epact	Concurrent	Lunar Cycle	Full Moon	Paschal Sunday	Lunar Age
532	10	Nulla	4	17	5 April	11 April	20
533	11	11	5	18	25 March	27 March	16
534	12	22	6	19	13 April	16 April	17

Table 1 – Excerpt from the Paschal Table Created by Dionysius Exiguus

The cryptic meanings of the headings used in this Paschal table are described in more detail below. Remember that when it came to keeping track of time, it fell on the shoulders of astronomers (back they were often called astrologers) to keep this information exact. Assuring the correct date for celebrating Paschal Sunday fell on the shoulders of astronomers, the trusted scientists of the time.

Table Heading	Meaning
Year	Calendar year number based on the Julian Calendar
Indiction	Roman-based 15-year cycle number (1-15) used in all legal documents
Epact	Day of the month on which a full moon occurs in that Julian Calendar Month
Concurrent	Day of the week (1-7)
Lunar Cycle	Number of the Lunar Circuit (1-19) for the full moon in the next column
Full Moon	Julian date of the previous full moon
Pascal Sunday	Mapped Gregorian Calendar Date
Lunar Age	Number of calendar days since the previous full moon

So figuring out when Easter Sunday (Paschal Sunday) occurred in the Gregorian calendar needed to account for:

- The ever-changing celebration date of Paschal Sunday
- The date of the nearest preceding full moon to Paschal Sunday
- The number of days after the preceding full moon on which Paschal Sunday was celebrated
- Whether a "leap day" occurred in that Julian calendar year

It is clear that the ancient Catholic Church was focused on knowing when to celebrate Paschal Sunday, one of the highest spiritual celebrations in the Church's calendar year. However, the focus on how much time passed since the birth of Jesus was not (another oops!). This gives you an insight as to how these ancients thought and what priorities they had…something to think about.

Dionysius Exiguus, the author of the Paschal Table, lived long before Pope Gregory XIII and was initially hired by Pope John I in 525CE to prepare a Christian basic chronology. Dionysius used the records of the time that accounted for the number of Easter celebrations conducted by preceding Popes to arrive at the correct number of years. But Dionysius made a simple math error (another oops accredited to the experts).

Dionysius' predecessor whose work was used to determine this date was a lesser-known pedantic named Theophilus of Alexandria. Theophilus created the second a set of tables based on the work of Patriarch Cyril of Alexandria around

390CE. So the key players in determining the correct calendar year for the currently-used Gregorian calendar were:

Date	Person	Lived About	Achievement
390	Cyril of Alexandria	376-444	Created first Paschal Table
440	Theophilus of Alexandria	unknown	Revised the Cyril Table
525	Dionesius Exiguus	470-543	Revised the Theophilus Table
1582	Pope Gregory XIII	1502-1585	Commissioned the Gregorian Calendar (the calendar in current use)

As another pedantic aside which may or may not have contributed to such errors is that what we take for granted today as clean drinking water was the exact opposite in ancient times. It was unknown why people would get sick from drinking water and so substitutes such as wine and beer were embraced instead. Even bathing was suspected as the source of these illnesses and so bathing was also greatly frowned upon. The fear of water was true at the time of Dionysius and the consumption of beer and wine were accepted as the norm. Perhaps these errors could be contributed to the fact that alcohol was routinely ingested as a substitute for water.

The impact of this fact is that the actual calendar year we all celebrate as today is not the actual number of years since the birth of Jesus. Now you see how this impacts understanding more about what appeared in the sky as the Star of Bethlehem when people had been searching astronomical databases for events that occurred in the wrong year (oops again!).

Whew! That was a mouthful! Now that you've struggled through the calculations and side stories, the thing to extract from this diatribe is this: our best estimates today place the birth of Jesus somewhere between 4 BCE and 7 BCE (it is common knowledge that Herod died in 4 BCE so Jesus must have been born before then). This has a huge impact on the people using the accepted calendar year to search for something occurring in the sky way back then: they are literally looking at the wrong period in time.

Now you see why it was important to understand a little about the evolution of the Gregorian calendar and the human errors made during its development. And it is also significant to observe that throughout history mistakes are made even by the finest minds of that time. These human errors are passed off as undisputed fact because of the titles associated with those doing the research (i.e., a pedantic). And it appears that this is indeed a trait of human nature in that the so-called experts still repeat this embarrassing fact of making mistakes in every walk of life (something I believe the human race will never outgrow).

So from what we understand today, Jesus was born <u>at least</u> 4 years earlier than Dionysius calculated and that makes the current calendar year at least 4 years older than stated (for example it is not year 2014, it is year 2018 and may even be year 2021).

An accurate range of dates for the birth of Jesus answers the first (and most important) question: when should we look in the sky for known significant astronomical events that could have been interpreted as the phenomenon called the Star of Bethlehem? The answer is this: sometime between 7 BCE and 4 BCE.

Look, Up in the Sky...

There are other questions that came along while answering my burning question of *What did the Magi see in the sky?* These other questions began to surface in 2012 with the end of the Mayan Calendar. Many folks believed that the end of this calendar predicted the end of the world and this would occur on December 21, 2012, at 11:11 coordinated universal time. But as you know, that didn't happen.

What caught my personal attention was a bit different from that of the media hype around December 21, 2011, that being the incredible length of time for which their calendar accounted. Their calendar "year" was also called the "universal cycle" or the "long count," a cycle of 2,880,000 (two million, eight hundred eighty thousand) days.

Although indirectly connected to my investigation of the appearance and subsequent disappearance of the Star of Bethlehem (where did it go?), I needed a serious attitude adjustment to change my way of thinking about the things that this event provided. In other words, I needed to learn to think "out of the box" better than I had ever before and this event helped me do just that. What the Mayan Calendar taught me was how to change my perception of time.

On 12-21-2012, the current long count (called the Pictun) ended. Counting backwards 2,880,000 days, the first day in this long count began on August 11, 3,114 BCE. However, the Pictun was <u>not</u> the longest unit used to measure time, not by a long shot. Much like the inch measures a tiny fraction of a mile, or a second the tiny fraction of a day, the Pictun (about 7,885 solar years) is a unit that measures a tiny fraction of time in much the larger Mayan calendar; other units in the Mayan calendar ticked off <u>huge</u> time periods. These are:

- Calabtun (20 Pictun or about 158,000 solar years)
- Kinchiltun (20 Calabtun or about 3 million years)
- Alautun (20 Kinchiltun or about 63 million years)

What captivated me was their enormous difference in attitude about time compared to any other known civilization. Today, we normally think of time immediately around us, like years, decades, and possibly centuries. Business plans routinely consider 5 and 10-year implications but the Maya thought far longer down the road than this. I cannot begin to imagine what the concept of time was like for a farmer raising corn next to a pyramid in Central America in the Gregorian year of 250 BCE. Countless generations preceded that farmer all with this same attitude about time so thinking about daily life within these enormous time spans appeared "normal." Much like knowing how old you are, these people must have known where their lives neatly fit into a tiny spec of their overall view time. They must have perceived their entire life as being like a second compared to a thousand years. Fascinating!

The length of time any culture existed can be inferred from the size of the units used to measure time in their calendar. The Maya have the largest sized calendar units on record, by a long shot. The size of the calendar units also hints at the origin of the calendar's introduction (the start date). To this start date add more time to develop the calendar's mechanics and the origin date of the culture is even older.

Again, counting backwards, this means that the Maya civilization and culture existed for over an unimaginably long period of time (before 63 million BCE). It also implies that a significant amount of time lapsed prior to this ancient date to develop the mechanics of this calendar's dynamics. In other words, someone did not just arbitrarily sit down one day and decide to set up the calendar in that way; this just did not randomly happen overnight. If this is true, then the Maya are the oldest continuous civilization in history, far beyond what is believed to be their age from surviving records.

The Maya achieved their highest state of development between the Gregorian calendar years of 250-900 CE. Along with their calendar, we know that the Maya placed great focus on mathematics and astronomy. In their astronomical studies, they developed a "grid" of background stars from which their celestial observations were made. This grid, similar to the twelve astrological constellations of the Zodiac used by astronomers and astrologers today, consisted of 13 Olmec constellations (365 days/year divided by 28-day intervals equals roughly 13 equidistant constellations).

The 13 Maya Constellations

During the study of the night sky, routinely wandering stars were identified moving through these background constellations. Today we call these wandering stars *planets* since they move against the background of all the other pinpoints of light (the word planet is derived from the Greek word *planomi* which means *wanderer*). It is these 13 Olmec constellations through which the planets and Moon move. Because of their wandering nature, these unusual non-fixed pinpoints of light were thought to have significance when compared to earth-based events.

The Maya did not hold exclusive rights to mapping these types of observations. All great empires studied the night sky, although not to the extent of the Maya. The Greeks, Romans, Chinese, Germans, French, just about every great civilization had and used astronomers. Each of these cultures developed their own set of constellations drawing imaginary pictures by connecting lines between stars. Movement within the stars was carefully measured and recorded.

It was unanimously believed by all of these cultures that the observations of atypical astronomical events influenced earth-based events. For example, the sudden appearance of a comet in the sky could be interpreted as an omen or an ally (something bad or good would happen) when compared to its coincidental position to other celestial objects. So the fact that a comet appeared was a puzzle in itself but the fact that it appeared coincidentally next to a star already believed to represent bounty or peril would be interpreted as meaning something quite different.

Taking this coincidence a step further, it was believed by all ancient highly-evolved civilizations that the movements of planets could influence the outcome of earth-based events. It was also believed that by properly understanding these positions, you could predict the outcome of a decision. This type of predictive analysis is today called *astrology*.

Once this word *astrologer* creeps into the study of the stars, people today think of these ancient scientists more as astrologers than as astronomers. People generally, and incorrectly, believe that since they did not have telescopes or a thorough understanding of the Cosmos, they were not serious scientists and possessed only the amateurish qualities associated with astrologers. This is the farthest thing from the truth. These were indeed serious people of science who dedicated their entire lives to the study of the night sky, who trained with others for decades to understand the mathematics of celestial behavior, and to impartially record events as they observed without bias or agenda. The degree of accuracy of the Mayan calendar is just one surviving testimony to the quality of their science. So let's get one thing clear: these astronomers called the wise men (or the Magi) were learned, dedicated, serious men of early science and not tricksters, fly-by-night amateurs, or charlatans. Like Galileo

Where, Oh Where did the Star of Bethlehem Go?

Galilei, Nicolaus Copernicus, or Albert Einstein these Magi were well-respected members of the scientific community. Below is a list of early astrophysicists and a brief summary of their contributions:

- **Andronicus of Cyrrhus** (100 BCE) – studied the sun, time, wind, and weather and placed the first weathercock atop the Tower of the Winds in Greece
- **Aristarcus of Samos** (230 BCE) – the first astronomer to mathematically prove the position of the sun at the center of our solar system
- **Eratothenes of Alexandria** (194 BCE) – invented geographic terminology still used today
- **Eudoxus of Cnidus** (347 BCE) – a Greek mathematician and astronomer who proposed the theory of proportions, a significant piece of Euclidean geometry
- **Gan De of China** (400 BCE) – compiled the first known star catalog
- **Hipparchus of Greece** (120 BCE) – compiled the second known star catalog and the founder of trigonometry
- **Kidnnu of Babylon** (400 BCE) – studied the movements of planets and the moon to such a degree of accuracy as to predict lunar and solar eclipses, the orbital period for one lunar cycle, and a highly accurate calendar
- **Naburimanni of Babylon** (500 BCE) – proposed the first theory of elliptical planetary orbits (ephemerides) and measured the length of a day and a solar year. His measurement of a lunar (synodic) month had a remarkable degree of accuracy (within 1.56 seconds!)
- **Thales of Miletus** (550 BCE) – the father of elementary geometry. Thales' work was used by Eukleides of Alexandria (aka Euclid of 300 BCE) who is considered the father of geometry.

This is a pretty impressive list of scientists, mathematicians, and astrophysicists and upon their shoulders other monumental achievements were made. These people were _really_ smart and while all of them understood astrology, they were not considered in any way to be mere astrologers. The wise men (Magi) were serious scientists on a search for an answer to a mysterious astronomical question.

The paths traced by planets against the background stars _usually_ but not always moved from west-to-east. Sometimes, these planets slowed their westward movement, stopped, and appeared to move backwards. This movement is today called retrograde movement. When any celestial behavior appeared to deviate from the norm, such a time was considered to be of greater significance than at others.

Eclipses (when one moves into the shadow of another) and conjunctions (when a planet is very close to another planet or bright star) were also considered to be events of interest.

Ancient astronomers were quickly thrust into the role of consultants who helped devise the timing for political decisions and military plans. Kings and rulers routinely consulted them to determine if the stars would reveal anything about an important choice that was about to be made. Today, people still consult astrologers (sometimes called psychics) when making choices about career changes, finances, love, or marriage. Newspapers carry columns based on the position of planets against the background stars and so it seems that this ancient art is still alive and well.

Finally we get back to how this all fits together with the Star of Bethlehem (I warned you about these long-winded explanations). So it seemed natural to me that at the time of Jesus birth any culture of any size would have these astronomers on staff and these astronomers would have great pull because of their direct contact with those in power and authority. In other words, if a celestial event were to occur that was known to happen but not be visible from their "home base," an astronomer would plead their case to the person in power. I imagine a dialog between an astronomer and a King being something like this.

Astronomer: In four months, there will be an astrological event of great significance, my lord.

King: And what would that be?

Astronomer: It is a solar eclipse, my Lord.

King: I see. I recall the last eclipse from which you accurately predicted our great famine.

Astronomer: Yes my Lord. You memory is excellent.

King: And why do you consult me with this?

Astronomer: With all respect, my Lord, I wish to observe it firsthand. However I must travel a great distance westward to do so.

King: I see. You need my blessing to make this journey.

Astronomer: Yes my Lord. Your understanding is correct.

King: You have my blessing. Take whoever you need with you and return. I am anxious to hear your findings.

Where, Oh Where did the Star of Bethlehem Go?

While the words in the actual dialog were undoubtedly different, the results were much the same. What most likely happened is that an unnamed ruler granted permission and funding for an astronomical expedition to observe an important event in the night sky that could not be observed from their current location.

If we are to believe that there actually was an astronomical event as recorded in the Bible commonly called the "Star of Bethlehem," that event would have occurred between the Gregorian calendar years of 4 BCE and 7 BCE. Also reported in the Bible, the wise men (the so called Magi) are said to have come from the east asking King Herod if he had seen this star. Let's look at what the Bible records regarding this astronomical event. Below is the translation from the New Revised Standard Version (NSRV) Bible of this record (see http://www.astronomynotes.com/history/bethlehem-star.html).

Matthew 2:1-2

> *In the time of King Herod, after Jesus was born in Bethlehem, wise men from the East came to Jerusalem, asking, "Where is the child who has been born king of the Jews? For we observed his star at its rising, and have come to pay him homage."*

What caught my attention in this segment of scripture was the celestial event that occurred, not the words used to describe the people involved. Herein hides the hint to the identification of the so called Star of Bethlehem. Shifting the focus from spiritual beliefs to scientific data, recorded in this passage are a few facts:

- King Herod was alive (this confirms the date of this meeting prior to Herod's death in 4 BCE)
- Wise men (aka Magi) met with this King
- Since the words "wise men" are used and not the words "wise man," we know there were at least two people who met with this King
- To be granted a personal audience with any ancient ruler, you must be someone important
- It is implied that these wise men originally saw a "star" from their "home" observatory in the east (past tense of the word "observed")
- These wise men scientifically interpreted the meaning of this star's unusual appearance to be astronomically and astrologically significant
- The astrological significance they reportedly assigned to it was that of a King being born
- This astronomical significance caused them to travel westward from their home observatory

Hmmm, the last fact is something to think about. Why would the Magi have to travel west? This is the hint I used to refine my focus as I will discuss in a subsequent chapter. The plot thickens so we have at least two things yet to uncover before understanding why the Magi's east-to-west travel was necessary:

- Which known viewable astrological or astronomical events occurred between 7 BCE and 4 BCE?
- Of these possible events, which would require the Magi to travel from their homeland to witness it?
- We know that the home observatory of the Magi was east of Bethlehem, but where could this be?

With my attitude of time adjusted from the Mayan calendar and their contribution to the science of astronomy, I used a combination of tools to understand the answer to these questions. Since all astronomers of this time studied the regularly-recurring stars and celestial objects to create these predictions, let's see what it was they found to be of scientific interest.

What Did Ancient Astronomer Do All Night?

Because of the corrected date for King Herod's Death, at least now we are looking up in the sky in the right timeframe to see what could be of interest to these ancient astronomers. We know that astronomers used the background stars called constellations as fixed reference points from which to observe and record any anomalous movement or appearance. And we know that the Bible recorded some hints about what the Star of Bethlehem was without disclosing a lot of detail.

We do know that, whatever it was, it attracted enough interest for at least two people consisting of at least one (and possibly more) astronomers collectively called the Magi to take a journey from east-to-west to learn more about this celestial phenomenon. So what in the world (or rather out of this world) was so interesting?

You have to put a few more things into perspective in order to understand their motivation. Unlike the night skies in cities today, there was no electricity, no neon signs, and no security lights. When the sun went down, it got dark, very dark. When the moon was not up, anyone huddled around a campfire could see thousands upon thousands of stars including the cloud-like band of light known as the Milky Way. As a comparison, from downtown Main Street in Tampa, Florida, you can see the moon and about a dozen stars on a clear night. But in a remote part of Siberia, you can see roughly what these ancient astronomers did – literally tens of thousands of stars.

Also remember that after sunset, there was no television, internet, radio, or iPods to entertain you. The only music you heard was someone singing, humming, or playing an acoustic instrument. Most folks went to bed when daylight disappeared and arose when daylight returned. Lamps and torches were used sparingly compared to the unthinking convenience of flipping a switch. Very few people stayed awake after the sun went down but those who did used the stars as another form of entertainment.

Imagine what jobs would require someone to stay up all night around the time of the birth of Jesus. With just about everyone asleep, your first thought may be of security guards or the military people or possibly bandits or criminals; you may also think of sailors and even those in authority to govern, and you would be right. Many of these night owls must have known the constellations of the night sky like the back of their hand, especially sailors. Once land disappeared the stars were the only source of direction to safely guide mariners across seemingly endless open seas. Imagine traveling merchants huddled around snapping campfires gazing up into the sky telling stories, checking weather, and verifying directions for the next day's journey.

Without streetlights to hide fainter stars, their collective dim glow provided essential information many relied upon for more than just fanciful predictions or casual observations.

And then there were the ancient astronomers who, just as they do today, stayed up all night long looking for signs of movement in these pinpoints of light or straining to see faint cloud-like smudges in constellations. And just as a sailor in order to tell directions would be intimately familiar with their numbers, locations, and positions within constellations, these ancient astronomers would faithfully record everything they noticed in massive journals. Not only did they record variations, they also confirmed positions. Through the duration of the night, they would sweep the entire sky. Much like a childhood game that compares two seemingly identical pictures to count how many things you can see what is different between them; a seasoned ancient astronomer would easily identify anything different in any part of the night sky from night to night. But unlike this childhood game where the more differences you find just makes you happy, when an ancient astronomer detected something different it was a very big deal that would involve the entire professional community.

But changes in stars were not the only thing an ancient astronomer fancied. While most of the stars stayed stationary, a few routinely wandered against these background dots of light. Over the decades and countless eons of time, the positions of these wandering stars were carefully mapped. From this recorded information, they could easily predict their movements far into the future.

It must have mystified these ancients most (if not all) of who did not understand the orbital relationship between the sun and the planets. What we take for granted today in knowing that the moon orbits the earth and the planets orbit the sun were misunderstood. From their point of reference of a dark field on earth, everything moved like a huge curtain with pinholes hiding a light behind it. Because these wandering stars did not fit this model, they were a mystery worthy of greater study. Without knowing basic astrophysics, everything overhead appeared to move in relationship to where one stood. For example, a typical Roman belief at the time of Jesus' birth was that the sun was pulled across the sky by someone driving a chariot.

These anomalous wandering stars were of special interest to these ancient astronomers since they seemed to violate the standard movement of all other stars. I can only imagine how many theories and speculations arose trying to explain these erratic behaviors. There is only so much information that can be obtained about anything when you do not understand even a tiny bit more about the whole picture. But these irregular movements still captured the attention and challenged the intellect of the ancients just as the wonder of a lunar eclipse still captures the attention and stimulates awe in the mind of a young child today.

Imagine for a moment that you are a seasoned astronomer training a student in the mechanics of these wandering stars. From night to night, you measure and record their positions against the fixed stars in constellations and over time of many weeks you see their minor movements revealed. In astronomy, sudden occurrences are extremely rare and it takes a detail-oriented person to make such patient observations of these tiny, minor movements. In a world today filled with instant gratification where with a click of a mouse button you can buy a hat made in Thailand and have it delivered to your door in a few days, patience on this level is completely foreign to most of us today.

But yet this is exactly what a professional ancient astronomer did night after night. It was the changes in behavior of these wandering stars that triggered speculation about its meaning. It was well documented that certain stars grouped together in constellations and these stars did not move. Every night, over a few consecutive weeks, they appeared exactly in the same places that they did on previous nights. But every now and then, one of these wandering stars would drift close to each other as if they were paying a visit. Sometimes, even two wandering stars drifted close to each other, something of greater interest than just one wandering star drifting towards another star that did not wander.

When one wandering star came very close to any other star, it was called a conjunction. Some conjunctions were far apart and others very close together. From the ground, it appeared that they *should* collide or crash into each other but they never did no matter how close they came together. This was yet another mystery that stretched their imaginations.

For example, if two chariots raced toward each other, one could pass safely by the other by allowing enough distance between them. But occasionally, one chariot got too close to another and they both would crash. This is what ancient astronomers expected to happen to these wandering stars in a conjunction and they anticipated that one day they would indeed crash together; they had no reason to expect otherwise with the limited knowledge they had. Imagine being able to boast about being the first astronomer to witness with your own eyes when this crash occurred. What prestige you would obtain, what notoriety within your elite groups of fellow astronomers you would achieve, and what honors would you receive? This may have been one of the reasons ancient astronomers were so intrigued by these conjunctions.

The other source of celestial mystery is called *retrograde*. Normally, all background stars in constellations moved slowly from east-to-west from night to night. Wandering stars moved slowly across these background stars from west-to-east. Wandering stars, however, moved slightly faster than the background constellation stars and these minor movements were also painstakingly recorded. But every now and then, the normal path these wandering stars traveled deviated from the otherwise routine west-to-east direction and appeared to slow, stop, and then reverse direction (they literally appeared to "backup"). When a wandering star slowed to a stop, the place within the sky (that is, in which constellation) where this occurred was interpreted as having a distinct meaning.

Certain stars within a constellation were assigned titles of great significance preserving the names of heroes or stories of battles. Entire legends ensued from these constellations where "stick figures" were imagined drawing out animals, people, instruments, and weapons. For example, in the constellation LEO one of the stars is named Regulus. *Regulus* is the Latin word for *Prince* or *little King* and in Persia this same star was associated with Persian royalty. The word LEO is Latin word for *lion*, a creature of great power and authority similar to that of a King. All cultures preserved their folklore in the night sky and typically one of these characteristic stars within these stick figures was thought to possess special powers or represent certain characteristics. One interpretation of a wandering star stopping at such an important star would be to associate that star's characteristic to this event.

For example, a star in the heel of a stick figure of a person could represent stability, agility, or strength. And a wandering star backing up near this heel-star would raise the eyebrows of these astronomers in developing their speculations. Imagine knowing from the records of those ancient astronomers who preceded you that such an event as the retrograde of a wandering star would occur. Witnessing that it did so (as predicted) would appear to others not trained in that science to literally be magic. It is one thing for a King to occasionally watch a wandering star move across the sky and quite another for this star to change direction. Logical minds would want to know *why* or at the least *what* such an event meant to his or her governance.

Ancient astronomers were most likely less focused on the prediction aspect of their trade (astrology) and more focused on understanding why such unusual behaviors occurred (the science of astronomy). However, since these Kings paid the bills, these ancient astronomers had to juggle both jobs as best they could. Just as orators became politicians or storytellers became actors, people with unusual patience and perseverance typically became astronomers.

The Magi possessed, if nothing else, these two characteristics as professional astronomers: patience and perseverance. They were concerned with far more than what the Bible discloses about their behavior. To me, they would be more concerned with the science behind the event.

Astronomers have their own language much like that of physicians, lawyers, and other highly specialized trades. To hear theoretical physicists argue about string theory or brain surgeons about stem cell treatments for Alzheimer's disease, most laymen would nod off in five minutes or less. However, getting any of these professionals at a birthday party or sporting event, they usually leave their specialized lingo behind and communicate with the same words everyday people do (use common language). Herein is a hint as to why the words used in describing the Star of Bethlehem to King Herod were chosen.

For an astronomer to ask any King if they had observed a conjunction or retrograde in the constellation Pisces assumes that the King is fully aware of these terms. Insulting such a King with improper language or assumptions, especially one that you just met, is something one just does not do. For such a faux pas, you could literally end up losing your life. It would be far safer for these astronomers to ask about the interpreted *sign* (what the King would be interested in) rather than the event (what the astronomer would be interested in). And it seems from this logic that this is precisely what the Magi did (use common language).

So the stage is set. A group of ancient astronomers observed something unusual in the patterns of the stars in the night sky from which they were intimately familiar, and it is this deviation from the norm that caused them to travel. When they arrived, they met with a King and used the language appropriate to the meeting out of respect for the office and the individual when inquiring about this star.

From the understanding that only significant changes in the night sky would be worthy of such an adventurous expedition, let's look next at the types of astronomical events that fit this criteria. In other words, what unusual event in the night sky would qualify for funding of such a lengthy expedition?

What Was So Important?

One night at his/her home observatory, an unnamed astronomer noticed something unusual in the night sky, something that was brand new or very different or even extraordinary. Whatever it was, this sighting "made the headlines" of the local news and before long, a band of at least two people found themselves traveling westward from their homeland in search of answers. At least one of these travelers was a professional astronomer who was most likely highly interested in the science behind the observed phenomenon and only moderately interested in its interpretation. In fact, the skill of interpreting the meaning behind this event may not have been that of the astronomer at all but rather the work of another with specialized talents. There is no way to tell for sure.

Once this band of travelers arrived near their destination, they met with a King and used respectful, non-scientific words (common language) to describe for what it was they were seeking. The dialog was intended to be disarming and superficial in an attempt to discover what this King's own astronomers may have uncovered. But what transpired from this conversation was unhelpful – from their perspective this King wished them well and sent them on their way. The true motive behind this was one of self preservation; this King hoped these strangers would unknowingly act as spies for his own devious purposes (human nature hard at work).

But let's turn our attention back to the discovery of this so-called star. What do we know about it? Not much at all; few words are mentioned. Putting on our detective hats again, let's see what else we can surmise from the evidence and what astronomical events we can uncover that occurred during this 7 BCE to 4 BCE timeframe. Surely, a few things may stand out and we will follow the clues just like a detective follows the money in a murder mystery. Let's see if there are other clues hidden in the length of time this star was observed.

Later in Matthew 2:9, after the meeting with King Herod, it is states:

> *The (ones) but having heard of the king went their way, and look! The star which they saw in the east went ahead of them.*

Although we do not thoroughly understand at this point what "...went ahead of them..." means, we can deduce that there was an element of surprise in this excerpt. These words are also ambiguous; it is unclear if the star was originally observed from their home observatory in the east or if it was observed in Jerusalem in an eastward direction. The latter seems unlikely since Bethlehem is due south of Jerusalem; this seems to confirm the former assertion (this star was first spotted from their home observatory). It appears from this element of surprise (*…and look!*) that at some time this star may have disappeared, reappeared, or changed brightness, so we have another valuable clue as to the behavior of this celestial phenomenon. What astronomical suspects meet these factual criteria? From here, we can start to eliminate those astronomical objects that do not meet all of our criteria.

So what are the possible celestial candidates before we start eliminating those unlikely suspects? What regularly-appearing (repetitive or predictable) events are considered to be of significance that an astronomer (or astrologer) consider atypical, unusual, or important? There are many candidates that fit these three criteria and they fall into the following categories:

- Comets
- Planetary conjunctions (alignments)
- Solstices and equinoxes
- Planetary retrogrades
- Eclipses
- Occultations (one celestial object passing in front of another)

Eliminating Unlikely Suspects

A combination of any of the above candidates would be considered to be even more significant than one alone. For example, if a conjunction occurred during an eclipse, that would be considered more significant than either candidate alone. But there are two other **irregularly**-appearing events far more difficult to scientifically prove:

- Nova or Super Nova (an exploding star)
- Meteor (a streak of light in the night sky)

Because of the Biblical reference to multiple sightings of this celestial object, some of these can be eliminated from the list since they are non-repeating one-time events or because the duration of that event is brief. We can immediately eliminate solstices and equinoxes, eclipses, occultations, and meteors. Planetary conjunctions by themselves, which only last for one to three days, can also be eliminated. Comets by themselves can be treated similar to conjunctions although they draw more attention since they are less-frequently observed. Both conjunctions and comets cannot be completely dismissed but their behavior must also coincide with another in order to make it an event worthy of a long journey. From this first-pass filtering, this leaves the following three possible celestial objects:

1. **Planetary retrogrades** (a planet pausing, reversing direction, pausing, and resuming normal direction)
2. **Comets** (long-duration events that brighten over time, disappear, and then return)
3. **Nova** or **Super Nova** (an exploding star)

This celestial object, whatever it was and wherever it appeared in the night sky, held significance associated with **royalty** (...*Where is the child who has been born king...*). That is, by its position in night the sky this celestial object would be associated with a King. The word "birth" could also be symbolically interpreted as that of a change in its appearance (brightening or coloring of an existing star, or even the appearance of a new star). Although this may or may not be the case, it is something worth remembering that may be of value in later investigations.

Those celestial objects historically associated with royalty are the stars **Aldebran** in the constellation Tarus, **Regulus** in the constellation Leo, **Antares** in the constellation Scorpius, and **Fomalhaut** in the constellation Pisces. Applying this association with "royalty" stars and constellations to our three celestial suspects, let's examine their behavior and see if they exhibit strong or weak characteristics.

Any **planetary retrograde** at the moment of changing direction (pausing) would be of greater interest than a planet in retrograde motion alone. When a planetary retrograde changes its direction of travel, it pauses for several days. If this pause occurred near a royal star, it would be a good candidate. Any **supernova** occurring in a constellation or better yet near a star associated with royalty would also be a suspicious candidate. Supernovae also remain bright for a while and then fade away and eventually disappear from view. However, a **comet** drifts through many constellations and although it is one that exhibits the newness or change characteristic, it is a less likely candidate because of its movement. One would not associate royalty with an astronomical object that moved away from the stars associated with that royalty.

This leaves us with only two candidates neither of which seem to be ideal: **planetary retrogrades** and **supernova**. This implies that we are missing something: *what have we overlooked?* Let's continue our investigation and see what circumstantial evidence we can uncover.

Where, Oh Where did the Star of Bethlehem Go?

The Usual Suspects

From these filters, let's look at what possible candidates there are from January 1, 7 BCE, through December 31, 4 BCE, as viewed from some place east of Jerusalem. Let's also see what major cities are east of Jerusalem (possible home observatory locations) and how long it would take to travel from those possible locations to Jerusalem. After all, the duration of time a celestial candidate appears must be longer than the time it takes to travel from east-to-west and it most likely took several weeks if not several months to make this journey.

Using the internet, I researched what others presumed to be the most likely celestial candidates. After all, I am not a fan of reinventing the wheel and if someone else already discovered what it was, then I would be happy. However, this was not the case and I pulled back from my pursuit of their possible candidates to consider more relevant or related issues. I suspected that some of these candidates were, like the old saying goes, "close, but no cigar." I felt like there was something everyone else was overlooking and I was determined to look "out of the box" at these candidates and see what that could be. If I was right, I could add another layer of filtering and narrow down the list of suspects even more.

As mentioned earlier, the two celestial candidates that meet the criteria vaguely described in the Biblical text are planetary retrogrades and a supernovae. These are the most likely candidates because they linger in the sky for a long time. We know for certain from the Biblical text that this star was implied to be first discovered from the home observatory somewhere in the east and also observed sometime after meeting with this King. However, *it is unclear if this star was observed continuously* from the moment of discovery, or that it appeared, disappeared, and then reappeared.

Again using the internet, I searched for all supernovae known reported by any other astronomers during this 3-year timeframe. However, I could find no reports of any nova or supernova during this period in history (although records are scarce that far back in time). Chinese astronomers are credited for recording the first observed supernova in 185 CE; however, this is disputed and may actually have been a comet. Since no scientific records survive, we cannot rule the so-called Star of Bethlehem as being a supernova but there is no way to prove or disprove it from recorded observations. For now, let's eliminate supernovae but we will come back to them later.

Well, that seemed simple, so I was left with only one option as a likely suspect: a planetary retrograde. Not only were these retrogrades well documented, but they are also repeatable since they involve the paths of planets that have been with us since the earliest recorded history. For example, a recent discovery of several cuneiform tablets dating 75CE reveals astronomical diaries whose translation is:

> *Month 10 [75 CE], the 1st of which will follow the 29th of the previous month. Jupiter, Venus, and Saturn in Sagittarius, Mars in Libra. On the 14th, Mercury will be visible for the first time in the east of Capricorn. On the 14th, moon sets after sunrise. On the 19th, Jupiter will reach Capricorn. On the 26th, Mars will reach Scorpion. On the 28th, last lunar visibility before sunrise.*

This ancient diary reads much like any modern scientific astronomical summary for the observable <u>lunar</u> and <u>planetary</u> movements of the month. Such tablets were commonly found in public libraries for anyone to read much like magazines are found in waiting rooms today. The minds of this time were intrigued by the events in the night sky and it was the job of astronomers to accurately record these events. If a retrograde were to appear, such an account would be found on a tablet similar to this one or in another type of reference book of the period.

The other hint as to *which* planetary retrograde was the one of interest to these early scientists was the fact that they needed to travel to observe it. This means that they most likely saw the retrograde <u>begin</u> from their home observatory but had to travel to observe its <u>end</u>. Normally, planets slowly move from west-to-east across the background stars of a constellation. At some point, this normal movement slows to a stop and then reverses direction (called retrograde motion). After several months, this retrograde motion slows to a stop and the planet resumes its normal west-to-east (prograde) motion.

With the naked eye (without the use of a telescope), a planet appears to stop for several days before changing direction and it is at this moment of stopping that correlations to nearby stars are made to associate its astrological significance. So there are two events associated with planetary retrogrades that the Magi scientists would desire to observe, that being both the starting of retrograde motion and its ending. If the start could be observed from the home observatory but the end could not, this would explain the need to travel from east-to-west.

At that moment, I started to uncontrollably jump up and down since I believed that this is exactly what happened. This is why travel was necessary. This is why they traveled from east-to-west as opposed to traveling from west-to-east. From their home observatory, the Magi could see this star enter retrograde but they couldn't see it exit. They had no choice but to travel. Sherlock Holmes would be proud!

This meant that the planetary retrograde observed by the Magi appeared above the eastern horizon before it disappeared months later below the western horizon as it exited retrograde. With their scientific mastery of knowing where and when such events would take place, it was a no-brainer for the Magi to understand how far west they needed to travel to observe its exit. Now, all that remained was to find a planetary retrograde that fit this last criterion: the dates on which they occurred.

I was a happy camper and felt convinced I was on the right trail. This theory all made sense, something I felt other theories did not. Now, all I needed to do was to identify what entry to a planetary retrograde could be observed east of Jerusalem across the desert whose exit could not. And of these planetary retrogrades, which occurred between January 1, 7 BCE, and December 31, 4 BCE. But before we go any further, let's see what a planetary retrograde is all about.

Backing up a bit more on this subject, there are a few things about the movement of stars in the night sky you should know before you can understand planetary retrograde. Unfortunately, this is where my Pedantic Mode again gets turned on but I believe it is important for you to understand a little about what these ancient astronomers found so fascinating about the night sky. I will be as brief and concise as I can.

The following images show normal star movement of one particular star. All stars behave just like this star. You can go outside on any night at 9:00 PM, **look due east**, and pick out any bright star to do the same. In this example, I will look due east from Tampa, Florida, a pick out the bright star named *Procyon*. It is shown below in the center of the circles in a group of fixed background stars called a constellation (we will talk more about constellations later). This particular constellation is named *Canis Minor*.

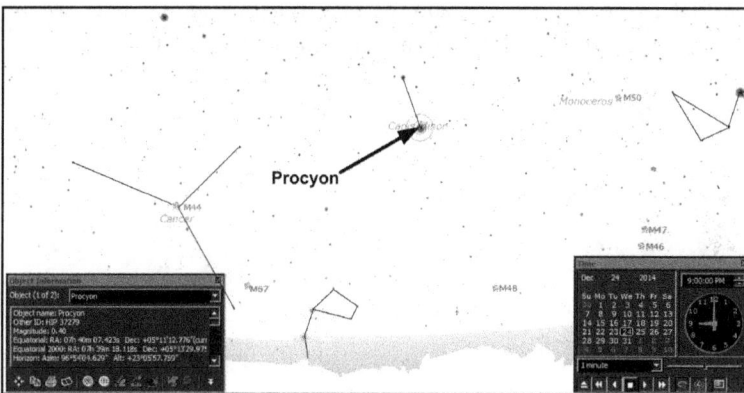

Looking Due East from Tampa, FL at 9PM on December 24, 2014

One week later, go outside again at 9:00 PM and look for this same star. It will appear somewhat higher above the horizon than it did a week before. Following this same example from Tampa, Florida at 9:00 PM, Procyon is a bit higher in the sky on December 31 than it was on December 24 (still in the center of the circles in the constellation Canis Minor).

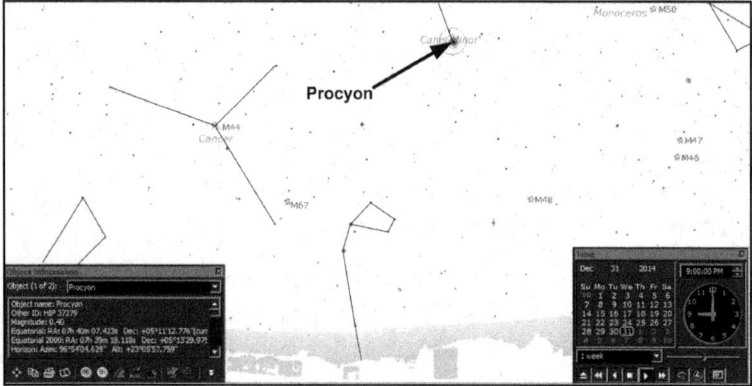

Looking Due East from Tampa, FL at 9PM on December 31, 2014

All stars, not just Procyon, make this gradual journey from the eastern horizon to the western horizon. Since they all move as a group with respect to each other, they are called **background stars**: they all move across the sky at the same rate. Planets, however, move slowly against these background stars. If you go out one week later on January 7, 2015, and look due east again at 9:00 PM, the planet Jupiter appears just above the eastern horizon. Following this same example from Tampa, Florida, at 9:00 PM, Procyon is even higher in the sky on January 7 than it was on December 31 (still in the center of the circles in the constellation Canis Minor). Here both the first sliver of the Moon and the planet Jupiter appear low on the horizon near each other. By the way, this closeness is called a **conjunction**.

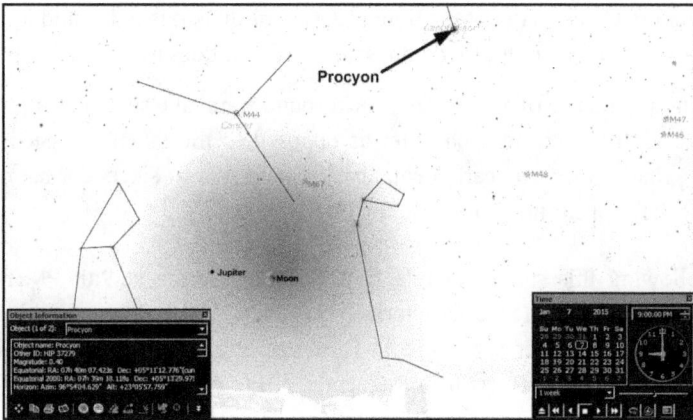

Looking Due East from Tampa, FL at 9PM on January 7, 2015

One week later, Procyon and Jupiter are even higher above the horizon, but the Moon has not yet come up. The Moon is making its normal west-to-east journey across the background stars and will appear about 11:00 PM above the eastern horizon. The Moon moves across the background stars much faster than planets because it is much closer.

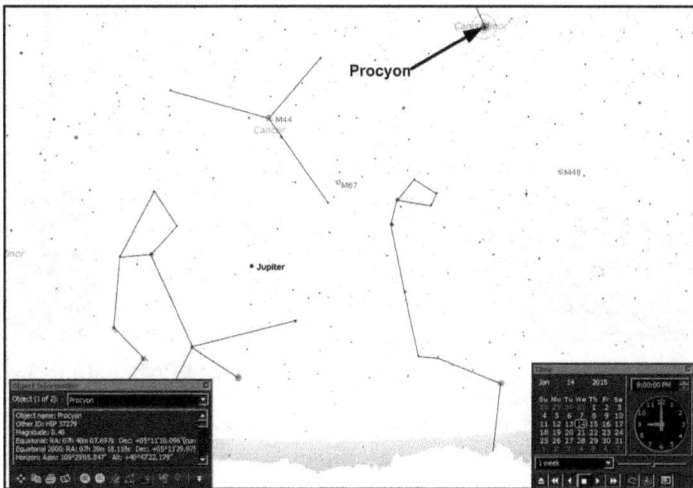

Looking Due East from Tampa, FL at 9PM on January 14, 2015

If you follow the position of Procyon over the next few months, it will arch overhead and eventually approach the western horizon. This is the normal east-to-west movement of all background stars. Procyon was in the same place in the constellation Canis Major at the time of Jesus' birth as it is today and it traveled from east-to-west back then in exactly the same way as it does now.

Without the use of a telescope, background stars take several centuries to appear to move with respect to each other as opposed to the far-more speedy moon or a planet. That's why we call them "background stars" since during one person's lifetime they appear not to move at all.

Following this same example from Tampa, Florida at 9:00 PM but this time looking **due west** at the western horizon on June 3, 2015, Procyon will be just above the horizon and Jupiter just above it. However, Jupiter will have moved away from (to the east of) Procyon (still in the center of the circles in the constellation Canis Minor).

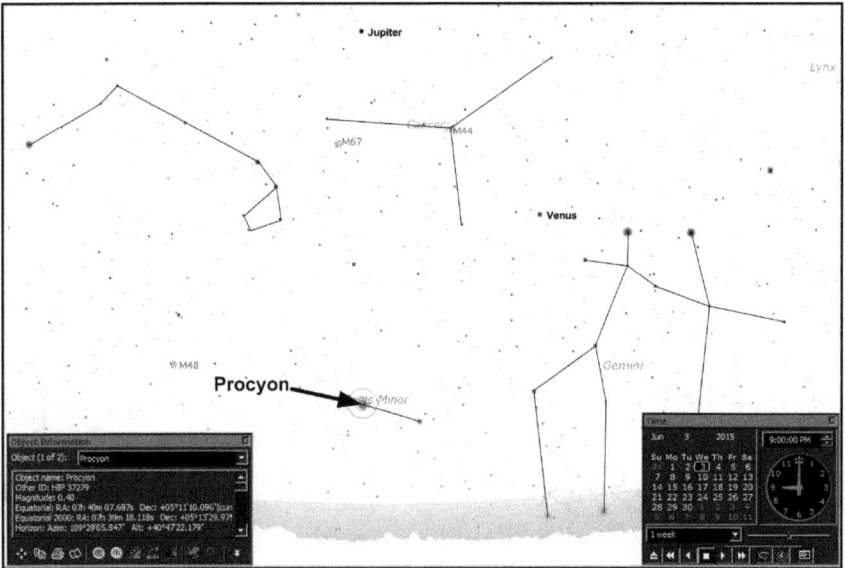

Looking Due West from Tampa, FL at 9PM on June 3, 2015

You can see that Jupiter has moved just a little over five months with respect to the background stars M67 and M44 as shown in the next image (north is up).

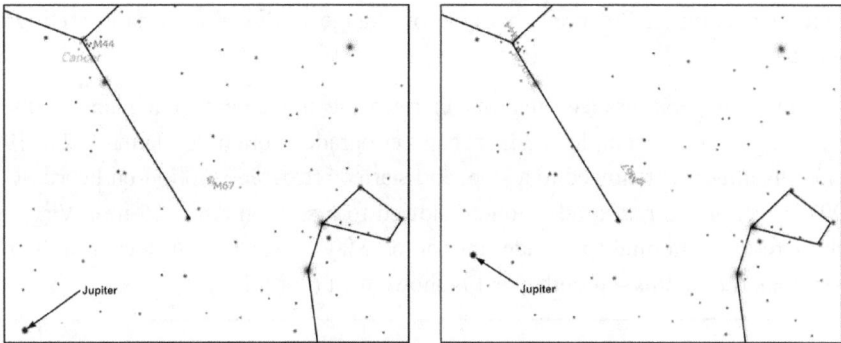

Jupiter on January 14 and June 3, 2015

As you can see, it takes a lot of patience and perseverance to map these normal, tiny movements of a planet. But this is exactly what ancient astronomers did in trying to scientifically unlock the mysteries behind their movements. Some planets move across the entire night sky and some only move up from the horizon, change direction, and then dive down below the horizon. Those that move across the entire night sky are called the "outer" planets and those that do not are called the "inner" planets.

So to quickly summarize, there are two types of stars that slowly march across the sky each night from east-to-west: 1) **stationary stars** grouped in stick figures called **constellations** and 2) **wandering stars** that move around in these constellations called **planets**. And you saw that both planets and stars move so slowly it takes about a **week** to really notice any of them moving at all. You also saw that over a period of a few **months**, planets *normally* appear to move almost imperceptibly slowly against those background stars in a west-to-east direction. But what is *abnormal* planetary movement? What is this called? Yup; you're right! This is called retrograde motion (sometimes called planetary retrograde).

What is Retrograde Motion?

Retrograde motion literally means *to back up or reverse direction*. Planetary retrograde means that a planet appears to reverse direction with respect to its normal west-to-east (prograde) motion against the background stars of a constellation. When plotting the movement of a planet, such as Jupiter, against these background stars, this minor movement can be traced. Without the use of a telescope, nightly changes are incredibly small and hard to detect so this gradual west-to-east movement is easier to see with the unaided eye when plotting changes from week-to-week. At some point, these plotted changes look like the planet is either looping around in a circle or making an S-curve. This type of change in direction is called retrograde motion.

But with perseverance anyone can trace the movement of a planet within a constellation. For example, Mars began retrograde motion on January 23, 1997, where its movement slowed to a stop, and started retrograde motion on February 12, 1997. It continued retrograde motion, slowed to a stop on April 25 near Virgo and finally resumed normal (prograde) motion on May 5 near Leo. A sketch of the path Mars took during this 4-month event is shown next (north is up).

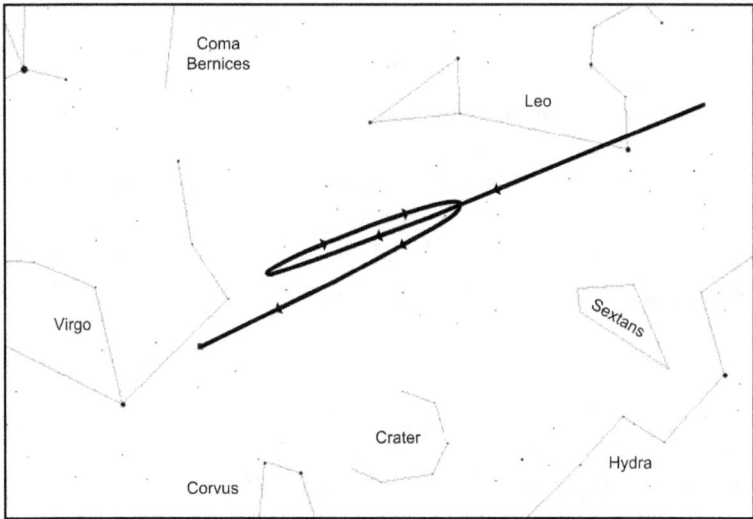

Retrograde Motion of Mars in 1997

Retrograde motion of all of the other planets is similar. Normal forward motion (prograde) is a very slow progression across background stars from west-to-east while all stars (including planets) move much more quickly in the opposite direction from east-to-west. The next image shows the path of the planet Jupiter during a recent retrograde. Each dot represents how far Jupiter moved against these background stars every 10 days. As you can see, movement as a planet approaches the entry or exit point, slows considerably (it appears to "stop" for about 30 days).

While this tiny motion is clearly evident through a telescope it is very different to see with the unaided eye. To the unaided eye a planet entering or exiting retrograde appears to literally stop moving for several days. This fact could explain other ambiguous Biblical verbiage (*...came to rest...* or in another translation *...stood above...*) describing the unusual behavior of an otherwise normally moving star (planet).

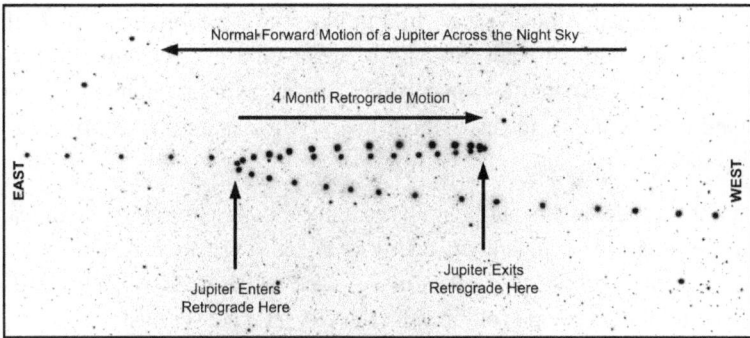

Retrograde Motion of Jupiter

Ancient instruments tracked the position of a planet during its retrograde and measured the angle of this planet with respect to the background stars. This angular measurement was then transferred onto a hand-drawn map of dots representing these background stars. The resultant drawing would appear much like the above time-lapse photography of the path of Jupiter's retrograde motion.

Retrograde motion is an optical illusion created by the movement of the earth with respect to the planet. One way to describe why this illusion appears is a phenomenon that photographers frequently encounter when taking interesting pictures of things in motion. For example, say you desire to take a picture of a moving car. If you move the camera to follow the car, the car will appear "in focus" and the background blurred. If you do not move the camera, the background is in focus and the car blurred. So it is with observing any planet against background stars where here your focus is on the background stars and not the planet.

Focusing on the Background Shows the Movement of the Car Lights;
Focusing on Background Stars Shows the Movement of a Planet

Congratulations! At this point you should have a crude understanding of the night sky and planetary movement. You are a typical representative of a "modern" society and it probably took you a while to ingest this information; some of you as smart as you are may still not thoroughly understand but you most likely got the general idea. Just imagine how long it took for ancient astronomers to thoroughly understand these same concepts without the advantages and education you have at your disposal.

From this crude understanding of astronomy, I hope you can relate at least just a little to the way these ancient astronomers thought of the stars and any unusual events they could observe. We can now get back to the detective work in this tale. Put on your thinking cap and roll up your sleeves; let's get serious!

Case Review

Good detectives examine all the evidence and eliminate suspects one by one. Does this person have an alibi? Does that person have a motive? And most important of all is to follow the money! Using similar elimination tactics, we have filtered down the list of all possible candidates leaving us with the remaining facts:

- The "home observatory" of the Magi was located somewhere east of Jerusalem (...*wise men from the East came to Jerusalem*...)
- The dates for this observation of a planetary retrograde fell between the years 7 BCE and 4 BCE (Herod died in 4 BCE)
- The list of all possible astronomical events – ignoring comets for the moment – has been whittled down to planetary retrogrades (lack of data recording the occurrence of supernovas during these three years plus no other astronomical events last long enough to be observed over a long period of time, that is, lasting long enough to travel from the East and still be visible while in Jerusalem).

All we have left to do is identify what planetary retrogrades occurred in that timeframe to understand which are possible candidates for the Star of Bethlehem. From this list, we can see which ones may have been more significant to observe than others. And remember, any astronomical event that coincides with any another astronomical event would be the most desirable to observe. For example, a planetary retrograde occurring during either a lunar or solar eclipse would be something an ancient astronomer would really be interested in observing. But first, let's see if we can eliminate other celestial suspects in our investigation.

It is safe to assume that these ancient astronomers did not have telescopes (duh!) and could only view planetary retrogrades they could see with their unaided eyes.

Where, Oh Where did the Star of Bethlehem Go?

Since Uranus, Neptune, and Pluto are too faint to see without a telescope (even way back then when there was no such thing as light pollution), only five planets – Saturn, Jupiter, Mars, Venus, and Mercury – remain that would have easily viewable retrogrades. That eliminates three more suspects (if you can't see them, you can't see follow them).

To filter down this list again, I applied my earlier assumption:

> *From their home observatory, the Magi could see this star enter retrograde but they couldn't see it exit. They had no choice but to travel.*

But where was this home observatory? Let's see what we can understand about where the point of origin (what city) from which the Magi may have come. From the facts, we know that:

- The Magi's home observatory was somewhere east of Jerusalem
- The Magi observed the entry to retrograde from their home observatory
- There is a huge Desert east of Jerusalem
- There are no large cities in this Desert that could support professional astronomers
- Some unknown King approved and funded the Magi's expedition to Jerusalem
- Planetary retrograde lasts about four months
- The Magi observed the exit of the retrograde from Jerusalem
- Lunar and planetary movements were recorded in great detail
- Cuneiform tablets provided the everyday person with common-language records of these lunar and planetary movements

What we can derive from this information is that the home observatory of the Magi had to be closer than four month's travel time away from Jerusalem, right? And it would make sense that they would allow for travel delays so as to not miss observing this event. So they would plan ahead and arrive long before this five-day event of exiting retrograde would occur. So where exactly did these scientists call home? We will find this out in the next chapter.

There's No Place like Home...

Now we have a good enough reason to travel for several months from east-to-west to witness a rare celestial event. This means that the Magi scientists would only risk traveling for about *three* months allowing them adequate time to find a good observation site and set up their instruments. If they arrived late, there would be no second chance. So what are the options? What cities are in kingdoms large enough to have a King and of these which are rich enough to support astronomers?

Below is a map of the geography east of Jerusalem. The huge Syrian Desert spans hundreds of miles between Jerusalem and the closest eastern cities. Of these cities, only a few could support a king or ruler that could afford to maintain his or her own astronomers and would be wealthy enough to fund an expedition to Jerusalem. In order of their distances from Jerusalem – as the crow flies –are Baghdad, Kuwait City, Basrah, and Persepolis.

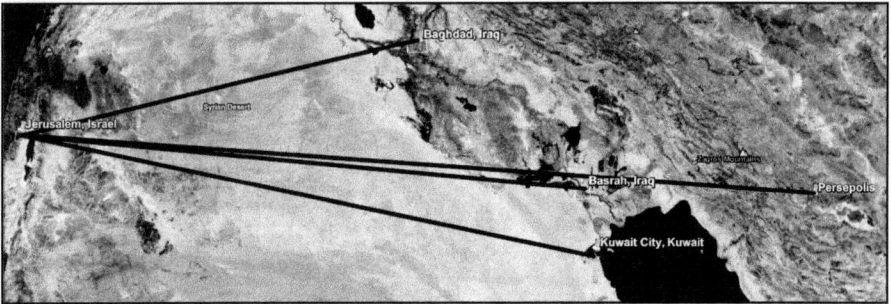

Large Cities and Geography East of Jerusalem

Baghdad (coordinates 33.3°N, 44.4°E) is on the well-traveled trade route between the Mediterranean Sea and China, today called the Silk Road. Travel along this route, while lengthy and somewhat perilous, is possible although not convenient for travelers of royalty or cast. **Kuwait City** (coordinates 29.4°N, 48.0°E) is on the water, a route more likely taken because of its relative comfort and speed. A quick trip around present-day Saudi Arabia could land aristocrats at the Gulf of Aqaba with a reasonably short land journey from there to Jerusalem. Nearby **Basrah** (coordinates 30.5°N, 47.8°E) nestles up against the Shatt al-Arab river that empties into the Persian Gulf. Travel from Basrah would take a route similar to that traveled from Kuwait City. The more distant **Persepolis** (coordinates 29.9°N, 52.9°E) supports several recently-excavated astronomical architectures (a science called *archeoastronomy*). German astronomer Wolfhard Schlosser confirmed the specific orientation of these buildings for the purpose of observing "the rising or setting of prominent celestial bodies on prominent dates of the year." Other archeoastronomy sites exist around the world for the sole purpose of similar observations.

We know that there were only two methods of travel at this time: by land or by sea. And we know that if the Magi were persons of cast, travel by sea would be preferred over travel by land. Although it does not preclude travel by land, travel by sea means that Baghdad and Persepolis – or any large land-locked city – is an unlikely candidate. This leaves us with either Kuwait City or Basrah, but which one? Again putting on our detective hats, let's investigate the cultures of these areas around 7 BCE-4 BCE and their focus on their spirituality and/or astronomy.

With a little more investigation, I discovered that of the two possible cities, Basrah was the better candidate since it was very close to Persepolis, the spiritual center for the culture of that time, but located in a warmer climate. Persepolis, at an elevation similar to that of Denver, Colorado, also housed official palaces for rulers of the time making this city a sort of summer retreat like "Camp David" of the US President today (see *Civilizations of Ancient Iraq*, Foster & Foster, page 145). But because of its extreme elevation and remote location, Persepolis was probably not the location of the Magi's home observatory.

Common sense dictates that serious astronomers would locate their home observatory somewhere in between Basrah and Persepolis far away from torches, campfires, and politics but closer to Basrah for easier to access equipment and supplies. Important to the location of any observatory is low humidity, light winds, and an unobstructed views of the horizon. Modern observatories are located on mountain tops for these same reasons.

Modern Astronomical Observatories Located in Hawaii at 13,500 feet
Photograph by Alan L
(http://www.flickr.com/photos/35188692@N00/2282306375)

The exact location of a key observatory in the Basrah area has yet to be excavated so I will use Basrah as the approximate location for this home observatory. A bustling city in 7 BCE, Basrah was part of the Charax region and under the control of the Parthians (early Persia) by King Phraates IV. The popular spirituality of that time was Zoroastrianism and the leaders from this region were widely known as the **Magians**. While the name is coincidental to the name associated with the wise men (Magi), this coincidence cannot be easily dismissed. Were these Magians actually the Magi? Perhaps…it does make sense but at this point it is still pure speculation. I will cover more about this coincidence later.

Now we must return to the original discovery that the Magi observed the entry to retrograde from their home observatory but could not observe the exit from this same observatory. Whatever this particular retrograde was, it most likely coincided with another astronomical event making this particular retrograde different. Otherwise, there would be no need to travel west to observe its exit.

I used TheSky$^{©}$ IV astronomy program to observe the sky from Basrah (coordinates 30.50°N, 47.83°E). I searched the skies for the entry to retrograde of any of these five candidate planets. I also used this same astronomy program to see if this same planet when exiting retrograde could only be viewed from Bethlehem (coordinates 31.7°N, 35.2°E). Much to my surprise, there were no candidates that matched all these criteria. This was puzzling to me because I felt I was on the right track. However, one of the candidates, Jupiter, showed something very interesting that I did not notice in any of the other retrogrades.

On August 19, 5 BCE, Jupiter **entered** retrograde. That night, the first glimpses of Jupiter from Basrah came at about 12:10 AM local time (just after midnight) when it rose above the **eastern** horizon. A picture of what the sky looked like at that moment in time as observed from Basrah is shown below. What caught my attention was how close M1 (aka NGC 1952) was to Jupiter (note that M35 is not a star). By the way, comet P2/Enke also appears very faintly in the sky due north of M1 as seen at the left edge of this image (above the letter "e" in the word "Object").

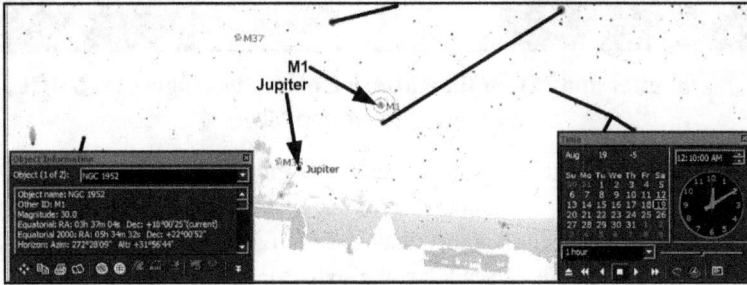

Jupiter and M1 on the Eastern Horizon on August 19, 5 BCE, as Viewed from Basrah

The next image changes the viewing position to Bethlehem to observe what this event looked like at the same orientation it appeared above the horizon in Basrah (12:10 AM local time). Remember, when you travel great distances from east-to-west, time zones change. When traveling from Basrah to Bethlehem you gain an hour just like when traveling from New York City to Chicago (your watch changes time from say 10:00 AM to 9:00 AM).

To see Jupiter and M1 at the same height in the sky above the horizon in Bethlehem as it was in Basrah, you would have to wait an hour (the hour gained by traveling east-to-west). Also note that by waiting this hour the orientation changes ever so slightly. Here P2/Enke appears above the letter "f" in the word "Information."

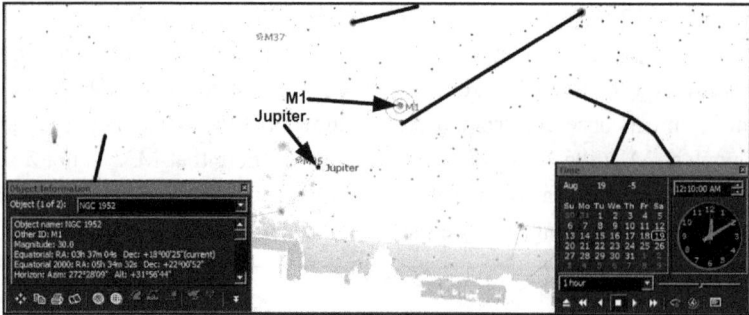

**Jupiter and M1 on the Eastern Horizon on August 19, 5 BCE,
as Viewed from Bethlehem**

120 days later, as Jupiter exits retrograde, it and M1 become aligned in what is called a *conjunction*. A conjunction is when two celestial objects are in close proximity to each other. Remember that in 5 BCE, M1 was a star about 9 times bigger than our sun. So let's see what the **exit** to retrograde looked like 120 days later on December 16, 5 BCE, just before Jupiter disappeared below the western horizon at these same two locations. The picture below shows the **western** sky from Basrah at that moment in time.

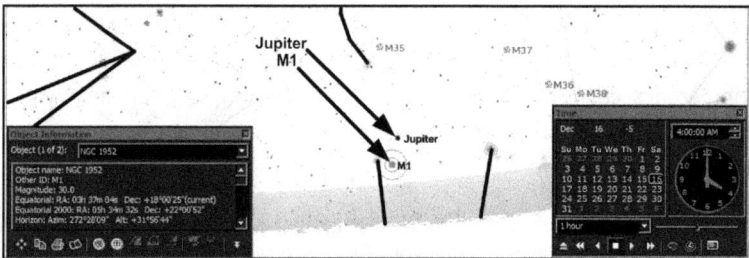

**Jupiter and M1 on the Western Horizon on December 16, 5 BCE,
as Viewed from Basrah**

Now jumping over to Bethlehem, the next image shows what Jupiter's exit from retrograde looked like from this location at precisely the same orientation just above the horizon (again time shown is local time).

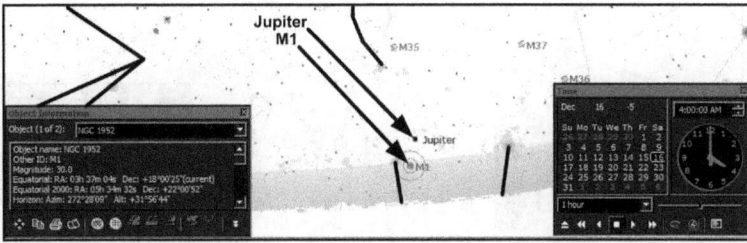

Jupiter and M1 on December 16, 5 BCE,
as Viewed from Bethlehem

This means that M1 would have been pretty bright and its conjunction with Jupiter interesting to any ancient astronomer, but an event in no way worthy of funding an expedition. Besides the exit from retrograde was easily visible from both locations; there was no need to travel west. Something else must have happened.

Just over three weeks later on January 8, 4 BCE, something else *did* happen: M1, Jupiter, and the Moon came into conjunction. Three celestial objects in conjunction is rare and this would be *almost* enough reason to travel west to observe it. At 2:30 AM, just before they set below the horizon, all three objects were in almost perfect horizontal alignment just above the **western** horizon.

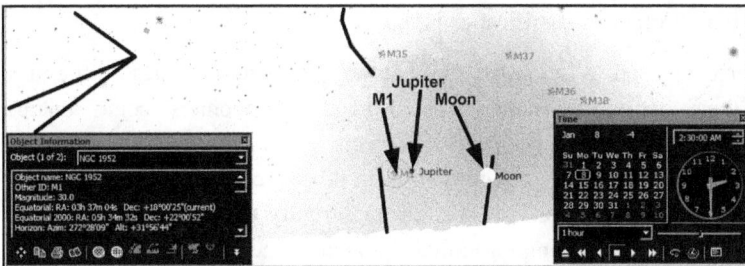

M1, Jupiter, and the Moon on January 8, 4 BCE,
as Viewed from Basrah

Remember that when traveling from east-to-west the time changes and with it the position of objects in the sky (just like comet P2/Enke did). When viewing this aligned conjunction from Bethlehem, the Moon is given one more hour to make its journey across the background stars (remember, the Moon moves west-to-east pretty quickly compared to the background stars). Below is what the western horizon was like as viewed from Bethlehem just before they set below the horizon. Note now that all three objects are in near perfect horizontal alignment as compared to their position as viewed from Basrah.

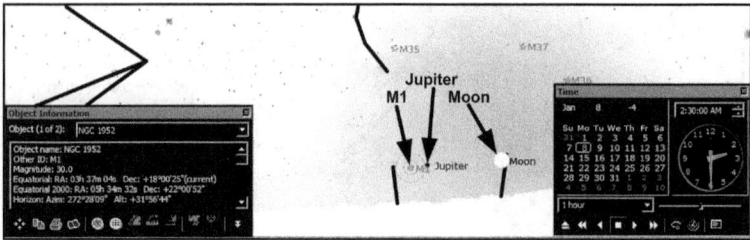

**M1, Jupiter, and the Moon on January 8, 4 BCE,
as Viewed from Bethlehem**

This near-perfect linear alignment is the astronomical equivalent of a smoking gun, fingerprints left at the scene, or the discovery of the murder weapon. These three objects in such a precise alignment would have been a good enough reason for any ancient astronomer to see it with their own eyes and convince any King to fund an expedition. Such an event would not repeat itself again for centuries and this is why these scientists were motivated enough to leave their home observatory and travel westward for several months to witness it with their own eyes. This was a truly once-in-a-lifetime event they just could not miss. And this event occurred in the constellation Tarus not far from the star Aldebaran, a star historically known to refer to royalty implying an astrological significance related to a King. We will talk more about this astrological significance later.

For the reasons of scientific study alone, if they traveled just a little further west from their home observatory, the Magian astronomers could make precise measurements of these three celestial objects. From the west, they could also see if the three objects in close proximity to each other would change the direction of any of their wandering paths. This was an opportunity they just could not pass up, no matter what else was going on in their kingdom. They needed to go west to observe this event but were these enough reasons for their King to fund this expedition?

Maybe; maybe not. But this is not the end of the story, not by any stretch of the imagination. This only gave the Magians a sound reason to travel to the eastern edge of the Mediterranean Sea. The fact that they were there at the same time of Jesus' birth was pure coincidence. However, this does not explain what else may have happened to cause M1 to appear above Bethlehem and why this particular star caught their attention. In the next chapter, we will investigate how this may have occurred and what it is they may have witnessed.

In the Beginning...

In 4 BCE there was an event occurring in the skies west of Bethlehem where the planet Jupiter on January 8 hovered low in the western sky coincidentally right next to the Moon (called a planetary *conjunction*). Right next to these two objects was the Crab Nebula (aka M1) and all three celestial objects, as viewed from Bethlehem, appeared in a straight line just before they disappeared below the western horizon.

| Conjunction in Basrah | Conjunction in Bethlehem |

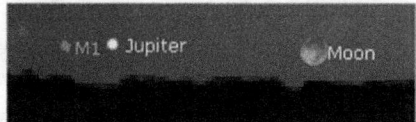

I felt that this conjunction was really close to warrant their King to fund an expedition lasting over eight months round trip but *still* not quite good enough. This would be a sizable investment to have a group of people just observe three things in a row low in the sky. There had to be something else and that something would have to be pretty spectacular. But what could it be? The final answer came from further investigations of M1.

M1 was documented as a supernova by Chinese astronomers on July 4, 1054. A French astronomer named **Charles Messier** assigned the M1 designation to this astronomical object in 1081, about 27 years after the Chinese observed its demise. Messier called this object a *planetary nebula*. Messier was actually searching for comets and developed a list of astronomical mudges confirmed not to be comets. He labeled these smudges as M1 through M110. Although a planetary nebula today, back in 5 BCE, M1 was a star called by modern astronomers a **progenitor star**.

I researched the behavior of the progenitor star of M1 by contacting Dr. Robert Stencel, the Director of the Astronomy and Physics Department at the University of Denver. In his professional opinion, here is what he had to say about what the behavior of the progenitor star of M1 may have been just before its demise:

> *The progenitor of the Crab nebula (M1) is thought to have been an 8 to 10 solar mass single, supergiant star – either a red supergiant or luminous blue variable – similar to that causing SN 1987A in the Large Magellanic Cloud.*
>
> *High mass progenitors seem to change their appearance on rapid timescales (hundreds or thousands of years) based on the best calculations available.*
>
> *Hence, it would be feasible to have looked different (color, maybe brightness) between BCE and AD times.*

Dr. Bob, as Dr. Stencil prefers to be called, explains that high-mass progenitors do not just suddenly explode; there is a period of time where this type of star changes its appearance. A change in either color or size is something ancient astronomers would have been able to easily observe without the use of a telescope. And *if* this change of the M1 progenitor was a thousand years before its discovery, this also puts its change in about the right time period (roughly the same time as the birth of Jesus).

Chinese astronomers on July 4, 1054, recorded a *supernova* event for M1's progenitor star, the term used to describe a star when it explodes. Today, the only thing that remains of this progenitor star that went supernova in 1054 is a glowing bunch of gas far out in space better known today as the *Crab Nebula*.

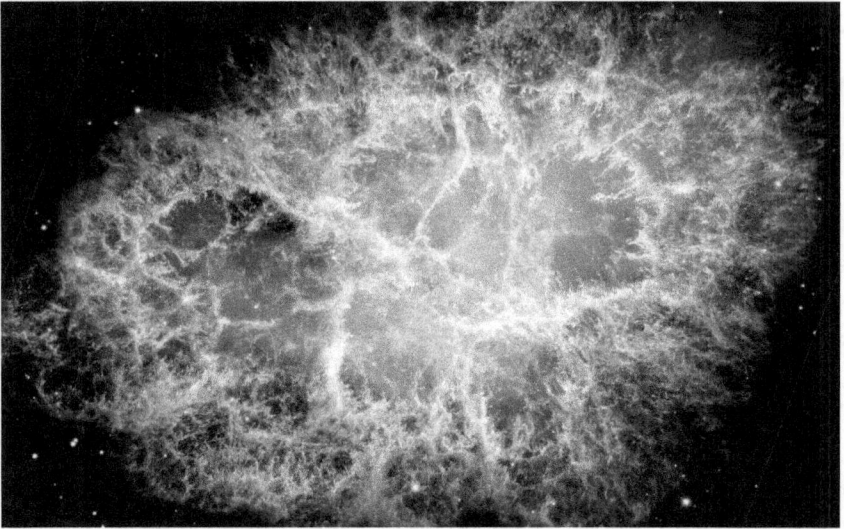

M1 Today (Hubble Space Telescope Photo Courtesy of NASA)

But way back in 5 BCE, M1's progenitor star was just an uninteresting **star** appearing much like any other star around it. Dr. Bob's explanation of the behavior of these progenitor stars confirms them to be <u>unstable</u> creating unusual brightening days, years, decades, and even centuries before their final demise as a supernova (see also *Nature* magazine, January 2014).

Here again we must put on our detective hats and create logical scenarios. What we know for certain about the M1 progenitor star is:

- The current celestial object called M1 encountered a supernova at sometime in the past

- Chinese astronomers report this supernova event as occurring in the year 1054 CE
- The size of M1's progenitor star was about nine times bigger than our sun (nine solar masses)
- Reverse-engineering the remaining evidence of M1 assigns the type of supernova its progenitor encountered as a **Type II supernova**

All stars eventually die in a supernova event but not all stars die in the same way. Depending upon their size (mass) and composition of elements, different things take place just prior to, during, and just after this explosion. These differences are grouped together and termed as the supernovae *Type*. The M1 progenitor is believed to have experienced a Type II supernova because of the remnants left behind (the bread crumbs leading a trail back for this assumption). When masses are high, such as the case for the M1 progenitor, Type II pre-supernova behavior is more likely to be unusual.

If you thought that the M1 progenitor just went "bang" with no sign of pre-supernova warning, you would be wrong. All of the remaining evidence indicates that this progenitor star experienced a weird death, one that would be better noticed by ancient astronomers and easily observed if their focus was on a nearby retrograde.

A Star's Relative Brightness

On a clear, dark, moonless night (under ideal conditions such as those routinely encountered in ancient days) a human eye can see about 2,000 stars overhead. If you stay up all night under ideal conditions, you can see closer to 4,000 stars as those in the west drop below the horizon and new ones appear above the horizon in the east. This is a lot to keep track of if you are an ancient astronomer.

Astronomers assign a *magnitude* number to star's relative brightness. Each magnitude number is 2½ times brighter than the one below it. Bright stars have low or negative numbers and faint stars have big numbers so a magnitude 1 star is about 100 times brighter than a magnitude 6 star.

Under ideal conditions, the faintest stars a human eye can detect are of magnitude 6½. A minor magnitude change of any single star could easily go unnoticed if you were not looking right at it or at something nearby. Even if you were looking directly at a familiar star, you would also have to be familiar with how bright that star normally would appear relative to those nearby to detect a change. However, if this change was huge such as a large change in magnitude and/or a distinct change in color, this would be more easily noticed.

Minor magnitude variations routinely occur in stars called *Cepheid Variables*. These stars experience regular variations much like waves in the ocean rise and fall changing the height of the sea. However, minor variations cannot be detected by a human eye. Any change in brightness or color would have to be pretty significant to be seen, such as that encountered during a supernova or a Type II pre-supernova (i.e., a Cepheid Variable also not a viable candidate for the Star of Bethlehem).

Summary

What present-day astronomers and astrophysicists believe to be correct about the behavior of the M1 progenitor star is:

- The color and/or magnitude of the class of star presumed to be the progenitor of M1 changed just prior to its demise
- This change in appearance can preclude the star's actual demise by decades or centuries

If the progenitor star of M1 brightened coincidentally on the date of Jupiter's conjunction with M1 and the Moon, ancient astronomers would have noticed since this conjunction is a normal event to observe in detail. Any change in the behavior of nearby stars would be taken seriously giving it a higher observing priority over anything else. And from these recent findings about the behavior of these progenitor stars, such brightening probably lasted several days or even weeks and then faded back to its normal intensity. But astronomers would make a point of going back and observing this star's behavior each night even if it returned back to its normal color and brightness. Patience and perseverance are engrained traits in any astronomer, even those of today.

We know that this change in appearance of the M1 progenitor could have occurred at the time of Jesus' birth. What if this pre-supernova behavior is what caught the attention of these Magi? Could this be what is called the "Star of Bethlehem?" I believe that this is true.

All of evidence I have uncovered points to what we know today as the Messier object M1 being the remnants of the Star of Bethlehem. All of the requirements of longevity, disappearance and then reappearance, changing color and/or brightness, and astrological appearance normally associated with leaders and Kings are present. Everything fits and the only wild assertion is that this progenitor star experienced a pre-supernova symptom coincidentally with the observation of Jupiter's retrograde.

Where, Oh Where did the Star of Bethlehem Go?

I suspect that M1's progenitor star brightened to an unusual intensity and may have also changed color as nearby Jupiter made its entry to retrograde. This change was easily noticed by the Magi during routine observations at their home observatory somewhere near Basrah. The Magi showed it to their King. who wanted an explanation to this phenomenon. The Magi – seizing the opportunity to do more science – explained that a 3-object alignment would occur soon after Jupiter's exit from retrograde and viewing only possible on the eastern shore of the Mediterranean Sea. The King agreed that this was a unique enough event that funding would be justified. And so, the planning for the trip began.

The change in M1's progenitor star would have been viewed by almost everyone in the kingdom and undoubtedly any other astronomer at any city observing Jupiter's entry to retrograde. This would have been the major topic of conversation around ancient water coolers. It is such widespread talk that would have made its inclusion in Biblical content for the sole purpose of reinforcing this date (linking one common-knowledge historical event with another). So it was important for the author of Matthew to not only include references to King Herod in the story about the birth of Jesus but also references to the Star of Bethlehem for the purpose of anchoring the exact time in which these documented events unfolded (memory by association).

But this is not the end of this story. The Biblical account in Matthew implies that the Magi met with King Herod in Jerusalem, something that does not quite fit for the position of the brightening of the M1 progenitor. But I suspect that this city could reference to the dominion of King Herod's authority. For example, references today are given to the capitol town of a state or the capitol city of a country with regards to their leaders. So when thinking about President of the United States, one immediately assumes a reference to Washington, D.C., the capitol of the United States. I suspect that – while they met with King Herod – the Magi's first encounter with this King was <u>not</u> in Jerusalem. I suspect that this reference was made to reinforce the time of this event rather than the actual location of the meeting just as references to King Herod associated the time of his reign with the time of Jesus' birth. But this too is pure speculation and we will see later how this plays out.

Jumping to conclusions in either direction (the Magi meeting with Herod in Jerusalem or the Magi meeting with Herod somewhere else) is presumptuous and cannot be taken as fact or even passing innuendo. Evidence to the contrary can reinforce one opinion over another and so our next chapter presents more bread crumbs to support or disprove my assertions.

One if by Land, Two if by Sea...

On the same night that Jupiter entered retrograde motion, the nearby progenitor star of M1 coincidentally changed color and/or brightness, and one alert scientist observed this event. This scientist – called a Magian – knew from historical accounts that in 120 days there would be a conjunction between Jupiter, M1, and the Moon.

From the accumulated data gathered through the centuries by his predecessors, this Magian also knew that these three celestial objects in this conjunction would be arranged in a straight line in a spectacular event. Also from his calculations, he knew he must travel west to observe this event with his own eyes and to do so he must secure permission and funding from the King, Phraates IV. After hearing the proposal and seeing the changed star with his own eyes, the King Phraates IV agreed to fund the expedition and insisted that an astrologer join these scientists in hopes of gleaning insight to future events. Now we have a convincing reason to travel for several months from east-to-west to witness a rare astronomical event.

♃

Traditional Astrological Symbol for Jupiter

Human nature dictates that King Phraates IV had different motives for funding this expedition than those of these scientists. Just like King Herod had ulterior motives for asking the Magi to share with him the location of the newborn king, King Phraates IV almost assuredly had similar reasons. People in power rarely invest in anything that does not improve their personal wealth, advance prestige, or assure ongoing power. People in power do not like to waste time or money and are typically dogmatically practical (rarely philanthropic). One thing is likely: when agreeing to fund this expedition, King Phraates IV hoped in some way to benefit from the knowledge obtained from his investment. What that benefit was remains a mystery.

Turning our attention for the moment back to the <u>astrological</u> significance of this event, M1 resides in the constellation Taurus (aka Taurus the Bull), a constellation long-associated with royalty. This sudden brightening because of its location could be astrologically interpreted as the birth of a strong new King. This would also explain why the Magi queried King Herod about a new King in their Biblically-recorded dialog. I will explain more about this later; let's return out attention for the search to the route.

If the location for the home observatory of the Magian scientists was either close to Basrah or the nearby spiritual city of Persepolis or somewhere in between, regardless of its location there were only two choices for a route of travel from which to choose:

1. **Overland**. This route would take our scientists on a very long and precarious journey around (via Baghdad) or through (via Dumah) one of the most desolate desserts in the world. Remember, these scientists were taking valuable instruments along with them and would want to assure their safe arrival at their destination. A lengthy overland route is almost certain to meet with disaster somewhere between their point of origin and their destination. Each night, these instruments would be unpacked from a beast of burden and then repacked the following morning. The chances of a mishap – a loose strap or unbalanced load – are high when traveling in such a way. Mishaps increase the chances for damage to valuable instruments or even loss of life. Traveling by land also meant a lot of walking and once they arrived at their destination, they would need time to regain their strength to do their research.

2. **By sea**. This route is far more likely since their cargo could be packed one time and safely stored below deck. It also makes sense that they would travel by sea since the Magi were members of the Aristocracy and would expect better accommodations than those available when traveling by land. Also, if the Magi left after Jupiter entered retrograde, they had less than 120 days to get to their destination, set up their instruments, and make their observations. Travel from Basrah by sea would take them around the coast of present-day Arabia and be a longer but much quicker route than traveling by land.

Ancient sources: Periplus of the Erythraean Sea, Pliny's Natural History, Strabo's Geography, Lucian's Navigium, Acts 27.

Key:
←——→ Combined products on land
Combined products on sea or river

Major Near Eastern trade routes in the Roman period, ca. 70 AD

500 miles

Water and Land Trade Routes of 4 BCE

Any expedition – either by land or by sea – would also include a small entourage of support personnel. People like cooks, animal managers, security personnel, guides, hunters, and a small maintenance staff are all likely candidates to come along. There is no way to predict exactly how many people were in this expedition but one thing is for certain: the Magi did not go it alone. They needed help and a lot of it for a journey this magnitude and duration.

So what route did they travel and how long did it take? Let's put on our detective hats again and see what we can come up with.

If you had to orchestrate and coordinate a trip of such duration, how would <u>you</u> do it? You are a smart person and you've already taken trips, probably several. What would you take along with you and what would you presume that you could obtain locally? Pause here for a moment and consider what it took for you to take <u>your</u> last trip. Even quick weekend jaunts require some thought and preparation; just multiply this planning effort (a 3-day outing) by a factor of 40 and then double that to account for the return trip. Remember, getting there is only half the journey; you also have to plan a safe return. What did you personally need to do, how long did it take you to arrange things, and what unexpected delays did you encounter despite your best-laid plans? This brief exercise should give you a glimpse as to what the Magi had to plan for their trip.

Obviously, the things the Magi must transport were their instruments, journals, star charts, and of course themselves. Money would also be needed to buy things along the way. Food and water were necessary to some degree, at least during open or unpopulated stretches. All of this planning takes time to arrange (called outfitting) and let's assume that it took the Magi a week to properly plan and stock these essential supplies.

If traveling by land, it would be about an 800-mile trip across the open Desert, or about 1,450 miles through Baghdad and along the Euphrates River. Traveling 8 miles a day, it would take about three months to cross the open Desert to reach Jerusalem. Traveling 10-12 miles a day, it would take about 4-5 months to travel by land via Baghdad to reach Jerusalem. The Baghdad land route puts the expedition at risk of missing the target observing date.

Traveling by sea, it would be about a 3,700 mile trip around present day Arabia. Traveling about 90-125 miles per day under reasonable conditions, it would take about 30-40 days to get to the northern tip of the Gulf of Aqaba and about another three weeks – including time to join a local caravan and outfit the expedition – to travel the remaining 170 miles to Jerusalem by land (about two months total travel time). So the time-critical choices eliminate the route through Baghdad leaving travel across the Desert or by ship. Which would you choose? By ship it is!

It seems that my first instinct to take a southern ocean voyage around present day Arabia is also the logical choice of routes. Let's assume that the expedition departed from the Persian Gulf and safely arrived on the northern-most tip of the Arabian Gulf 6-weeks later at the city of Aela (present day Aqaba). The express purpose for landing here was to connect with local caravans that traveled the southern route of the Silk Road (the safest way to travel) to the Mediterranean Sea. At Aela they had about a month left before Jupiter exited retrograde and another two weeks before the conjunction between M1's progenitor star, Jupiter, and the Moon (plenty of time). From Aela, the Silk Road led north to the Mediterranean Sea via Jerusalem. It is less than 200 miles from Aela to Jerusalem and that meant they had to make an average of 7 miles per day (easily doable with plenty of time to recover from mishaps).

The southern Silk Road trade route taken from Aela to Jerusalem skirts the western edge of the Arabian Desert following the eastern side of a 2,000-foot high mountain range. This mountain range blocks direct access to the Mediterranean Sea creating a wall 180 miles long. About 60 miles from Aela is a thriving and well-frequented caravan stop called Petra. Although in the middle of a desert, recent excavations have uncovered a labyrinth of water storage caverns. It seems that these ancient residents harvested rain that seeped through the porous stone, stored it in these caverns, and sold it to the travelers of this route.

Petra Amphitheatre where Travelers Relaxed

From Petra, it is another 60 miles to a mountain pass that finally exits the desert to a town on the western side of this mountain range called Feifa. Pushing another 15 miles west lands them in a town called Ein Tamar. From there, the caravan would travel 144 miles to the southern tip of the Dead Sea where it would then wind 35 miles more along its west bank (still with this wall of mountains on their left) eventually reaching the town of Almog and finally connecting to the main road to Jerusalem. Once in the safety of Jerusalem, the expedition could restock supplies and make necessary repairs before the final leg of their journey.

Caravan Route from Aqaba to Jerusalem

At this time in history, Jerusalem is a bustling hub of trade for goods destined for Greece and Rome. The new seaport at Sebastos built by King Herod to spur this trade lies just to the west. In Jerusalem, the Magi could rest after such a long and arduous journey and plan the best location for viewing the event. Of course dark skies, low humidity, good weather, good elevation, and safety are all considerations for choosing a site from which they would observe. With its activity, viewing from Jerusalem would not be a good choice. However, Bethlehem is only 4 miles to the south and also offers higher elevations with expansive west views and less humid conditions, all more favorable for optimal observations of this event.

This expedition and its entourage would gain attention as soon as they arrived anywhere. While in Jerusalem King Herod undoubtedly heard of their caravan's arrival through his own network of intelligence. Like any politician, he would request an audience with representatives of any foreign country to establish some form of diplomatic relations. Paranoid politics back then were no different than they are today and from such a meeting the true purpose of the Magi's expedition could be determined, hopefully as a non-threat.

This meeting is summarized Biblically, so let's see what insights we can glean from the text. Remember, in all political discussions, there are layers of words that appear to mean one thing and in fact mean something entirely different. Savvy ambassadors are aware of these techniques and routinely banter them back and forth. Being close to King Phraates IV, the Magi would have been exposed to some of these skills and be somewhat prepared for such engaging dialog. Although they were scientists and not politicians, they did their best by trying to assure that their presence was not to spy on the King or the country, but instead to do research on an astronomical event with obvious astrological significance. Below is this text as recorded in Matthew 2:2-3:

> "Where is the child who has been born king of the Jews? For we observed his star at its rising, and have come to pay him homage."
>
> When King Herod heard this, he was frightened…

Remember this:

1. The reason these astronomers (scientists) traveled all that way was not to see if a new King was born but rather to observe a celestial event.

2. An astrologer *would* say such things so this meeting was most likely conducted been between the group's astrologer (using common language) and the King.

3. Also remember that all foreign travelers were initially viewed as spies, something of which these Magi would be aware. Making such a statement implying the overthrow of a reigning King would not only be unwise but also could cause them to lose their lives.

My suspicions are that this dialog was altered from its original content. It would make more sense for the Magi to avoid making threatening references to a foreign King that would put their own lives in peril. On the other hand, putting dignitaries from a foreign government could undoubtedly incite a war.

Let us take the text as stated but know one thing is clear: after hearing this King Herod became even more paranoid than just thinking the Magi may have been spies from another country looking for opportunities to invade. The problem seemed to be more serious than he originally surmised and his reaction later in the text of Matthew underlines his paranoia. First let's summarize what we know as Biblically reported.

- The Magi appeared in Jerusalem unannounced
- The Magi are from a foreign country
- King Herod is the ruler of the territory around Jerusalem
- King Herod is responsible to Rome for maintaining the peace in his territory
- It is unclear to King Herod as to what purpose the Magi have for being in his territory
- King Herod suspects the Magi are spies and calls a meeting with them
- King Herod is a crafty politician
- The results of this meeting with the Magi make King Herod more fearful than he originally was

Where, Oh Where did the Star of Bethlehem Go?

Herod now fears something very personal: an overthrow of his authority. As detectives, we must now put on our thinking caps and try to develop logical scenarios that would fit human nature in a panic. Below is one scenario that seems to fit the evidence.

Being the cunning politician that he is, King Herod uses doubletalk common in such discussions to leverage as much of an advantage to the situation as possible. Although appearing to be a continuous meeting, what transpired was actually not. In the first meeting, King Herod discovers that these Magi have found a sign in the heavens indicating the birth of a King. I suspect that there was some confusion involved in the interpretation of these words (after all, the Magi spoke Persian and the King most likely did not and an interpreter was employed in these meetings).

The Magi explained to the interpreter that they observed a star that suddenly brightened in a constellation associated with royalty. This interpreter, no doubt expecting doubletalk to be delivered during these meetings, was trying to "read between the lines" and understand exactly what was being said by the Magi. Listening to his interpreters, King Herod took the translation to mean that this star implied a personal threat. From this counsel, King Herod had to regroup and think about the implications of this meeting. This required more information to eliminate the possibility that the translation was wrong or that his reading the doublespeak was wrong (as a seasoned politician, Herod would not have jumped to conclusions).

After some uncomfortable pleasantries most likely including discussions about where would be the best place to do astronomical observations, King Herod probably dismissed the Magi to do his own research. King Herod had another meeting where he "...called together all the Chief Priests and Scribes..." to confirm or disprove his fears. The consensus of this second meeting is that King Herod's suspicions were indeed justified since a prophecy existed stating a child would be in Bethlehem that would become a King of Israel. With his suspicions confirmed, King Herod now puts a plan together to use the Magi as unwitting spies to locate this suspected newborn. King Herod now has enough information to arrange another meeting with the Magi to execute his devious plan.

Any King receiving information even about an unconfirmed threat to his position would not treat it casually and would not leave strangers in his country unmonitored; he would want to "keep an eye on them." Herod most likely proposed one of several options: an escort to Bethlehem, offered the Magi to be joined by his own astronomers, or some other disguised covert operation so that he could keep tabs on their movements. Runners could then easily convey any important messages over the short 4-mile distance between Jerusalem and Bethlehem and keep the King immediately informed of any developments.

Reconvening with the Magi, King Herod then tries to use the presence of these scientists to covertly flush out the location of this child threat. He would recommend a nearby dark viewing site on a hillside just outside of Bethlehem where they could set up for the upcoming event (what he really intended was to put the Magi in the vicinity of the prophecy to make it easier for them to find this child).

So it seems that King Herod would succeed in his devious plan to find the child threat by using the Magi as unwitting spies. But I also suspect that the Magi were a bit savvier than the King gave them credit. I suspect the Magi explained they were there to observe Jupiter exiting retrograde but not tell him about the following conjunction. I suspect that they were wise enough to see through part of this plotting and scheming. I can only imagine what the Magi must have thought along the route to Bethlehem; the tension in the air from this meeting must have made their skin crawl. Now the Magi had to set aside all these feelings and focus on what they came all that way to do: science.

Twinkle, Twinkle, Little Star...

Before we continue, let's review from a less detailed level what events took place that led the Magi to Jerusalem. Looking down from an altitude of 10,000 feet, this is roughly what happened:

On August 19, 5 BCE, just after midnight, a Persian astronomer was busy monitoring Jupiter entering retrograde from his home observatory. With state-of-the-art scientific instruments in hand, he recorded in precise detail how it behaved noting especially its current and predicted position. Out of the corner of his eye, something magical happened: a star in the constellation Taurus, one he had seen thousands of times before, suddenly became much brighter and/or changed color. Realizing immediately that this unique event must be shared with the King, he started to review its associated astrological implications.

He studied the charts left by his predecessors and projected forward in time where the position of Jupiter would be when it exited retrograde. He then analyzed the associated astrology for that position knowing the King would be more interested in its ability to foretell the future as opposed to its scientific implications. After gathering the information he knew the King would require, he looked further ahead into the projections and discovered an even more amazing astronomical event would take place a few weeks later: a linear conjunction between Jupiter, the Moon, and this bright star.

For this once-in-a-lifetime event, this Persian astronomer convinced the King to fund an expedition to observe the exit of Jupiter's retrograde and the subsequent 3-object conjunction. He leveraged the astrological "royal" importance of this star in Taurus to get what he wanted from the King. It would take months traveling by ship and land to reach a point far enough to the west and still reach it in time to do the science he desired.

Outfitting his expedition, a company of scientists and support staff departed by ship and took it as far west as it could go. They then traveled the remaining distance by land following established trade routes for safety. After leaving the Persian Gulf, this expedition rounded the southern tip of Arabia, sailed up the Gulf, and landed in Aqaba. There, they discovered an impassable 2,500-foot tall wall of mountains blocking a direct route to the Mediterranean Sea and also obstructing any hopes of observing this event. They had no choice but to join a caravan headed northwest for Jerusalem, the safest and shortest route to the western side of these mountains.

Once in Jerusalem, this expedition's arrival gained attention by the local politician, King Herod, who was highly suspicious of their presence and initially suspected this expedition as being spies. After consulting with his staff, Herod found that they indeed were not spies but had uncovered something far more perilous: the prophesied birth of a child who would grow to be a powerful King. All Kings at this time are paranoid politicians and hearsay of any threat to their authority – even that of a newborn child – would be taken very seriously. Herod, in typical Roman reaction, would have no choice but to eliminate this presumed threat before it had a chance to materialize: Herod would have to find and kill this child.

Scrambling to find a devious plan King Herod hatched a plot that would unknowingly leverage the expedition's honorable intentions against his personal desire for self-preservation. It is Biblically recorded in Matthew 2:8 that King Herod "…sent them to Bethlehem…" to observe the event, the place most likely for them to succeed in locating this so-called child-king, and asked them to share what they learned. The expedition then departed most likely under a discretely watchful eye.

During the journey to Bethlehem, there is another hint in Matthew 2:9-10 about this star:

> …*When they had heard the king, they set out; and there, ahead of them, went the star that they had seen in the East (or at its rising)* until it stopped over the place where the child was…

This translation is harder to interpret – or at least interpret from an unbiased perspective. It does confirm that the phrase "at its rising" meant that the Magi first spotted a star and that it grew brighter. The only surviving original Greek account reads a little different:

> …*went ahead of them, until having come it stood above where was the young child*…

My personal common-sense interpretation of these words is this: After the second meeting with King Herod, the expedition headed south to Bethlehem. During this short trip, the Magi saw this same star they first observed from their home observatory. Shortly after this sighting, the Magi arrived in the sleepy little town of Bethlehem, the same town where the child was born (and still resided).

After all of the planning, traveling by sea, traveling by land, and finally an unplanned run-in with a paranoid politician, the Magi finally stood where they envisioned so many months ago. I can only imagine how relieved they must have felt knowing they had arrived with plenty of time to spare to do their observations. There must have been quite a celebration that day in Bethlehem.

After this celebration, the first decision facing the Magi was *where to conduct their observations*? Astronomical observations performed "in the field" are very similar to those performed from a stationary observatory. Their minimum "in the field" site requirements for any serious observation would be:

1. An open and flat area so their instruments could be easily used and leveled
2. A treeless place with unobstructed views of the western sky
3. A dark and remote location so torches and lamps would not harm their night vision
4. An area high above any valley so as to avoid ground fog or excessive humidity

The prerequisites for astronomical sites haven't changed much in the past 2,000 years. These are the same reasons why professional telescopes built today are located atop high mountains in dry regions far from any city lights (see the background behind the decision for the location of the Alma Observatory on the Chajnantor plateau, 5000 meters above sea level in northern Chile). From a rooftop in Bethlehem, they probably were able to spot several likely locations.

For "in the field" astronomy, an open field with low-growing grass similar to that used for grazing livestock would be a good choice. A high plateau would be an optimum choice with easy access for animals carrying equipment and a western edge falling away toward a valley.

The Biblical account of Matthew gives us another hint in as to where the Magi set up their instruments. Here the phrase "came to rest" implies "setting" just as the Moon sets in the west. If the Magi were to observe this conjunction setting over Bethlehem, they would have to view the event from the *opposite* direction, somewhere east of the town. There were broad grazing fields in this direction and viewing from these fields would cause Bethlehem to appear below this astronomical event. For the moment, let's assume this is roughly where the Magi set up their observing equipment and see where this leads.

The Shepherd's Field

There is a mountain trail about two miles southeast of Bethlehem with a clear view of the western horizon, an ideal place for such an observation. This hilltop, near the present-day city of Beit Sahour (a.k.a. Bayt Sahur), provided a reasonable elevation – above the humidity in the valley below – from which this celestial event could be observed. Here, it would seem reasonable that the Magi camped alongside whatever livestock was already grazing there – including those who watched over them (shepherds) – all waiting patiently for Jupiter to exit its retrograde motion. I suspect that this field commonly called the "Shepherd's Field" is the same place the Magi observed this event.

There is a lot of disagreement about where this so-called Shepherd's Field is located and it is not Biblically confirmed that the two groups camped out anywhere near each other. However, again putting on my detective's hat, I investigated this possibility. Recall the earlier image of this event as synthesized by the astronomy program TheSky VI:

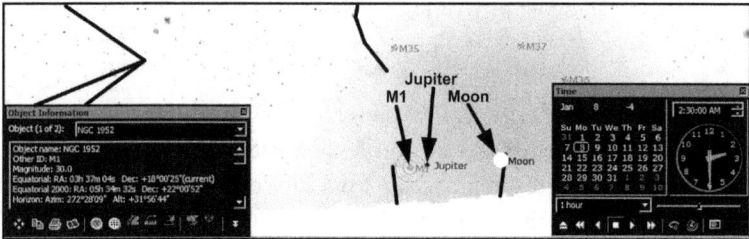

**M1, Jupiter, and the Moon on January 8, 4 BCE,
as Viewed from Bethlehem**

The information provided by the program in this image not only shows a picture of this event but also the data for the angular position of M1 (the Azimuth and Altitude angles). The Azimuth angle is a clock-wise measurement from due north. It is then a simple task to draw a line at this angle through Bethlehem to show where on the western horizon M1 would have appeared (272.5-270 is 2.5 or about 2½ degrees north of due west). Drawing a line in the opposite direction would help locate where the Magi may have stood to see M1 directly above the town. Following this line, you could see what hillsides fall on its path that would be more favorable for viewing this event (from a higher vantage point). Below is an aerial view of Bethlehem with a line drawn towards the direction of M1; this same line is also extended in the opposite direction.

**Direction of M1 on January 8, 4 BCE,
as Viewed from Coordinates 31.7°N, 35.2°E**

All we know for sure at this point are two things:

1. This is the general direction from which M1 could be viewed
2. These are the current coordinates for the heart of ancient Bethlehem

It is Biblically recorded that King Herod sent the Magi to the town of Bethlehem and this town is built on a 2,500-foot high ridge in the southern portion of the Judean Mountains. The surrounding land averages between 1,800 and 2,200 feet in elevation. The Ottoman tax record and census from 1596 indicates that Bethlehem had a population of 1,435 and current estimates are close to 30,000 so the "old city" as it is known today is most likely all that existed at the time of Jesus' birth. In other words, back then Bethlehem was a <u>much</u> smaller town than it is today, nowhere near as sprawling with a lot of rural communities.

Moving the coordinates for the heart of Bethlehem of course changes the perspective anywhere along that line. So, if we expand the width of the line by a very generous 1,000 feet in each direction (2,000 feet end-to-end width), it gives us a better idea where the Magi could have chosen a viewing location and still have had the conjunction appear just above this town.

**Possible Viewing Areas of M1 on January 8, 4 BCE,
as Viewed from Coordinates 31.7°N, 35.2°E**

If the Magi set up on a hillside east of town, and if the progenitor of M1 flared just prior to setting, this conjunction would be see as setting on top of the town. The bright progenitor would eventually disappear behind the distant buildings (due to the natural east-to-west movement of all stars as described earlier) and could be interpreted as it states in the Biblical text as "...stopped over the place..." If you were someone not trained in documenting astronomical events, these words may indeed be what you might use to describe what happened (using common language). Regardless, let's keep going.

My additional investigations uncovered another astonishing coincidence. Helena, the mother of Roman Emperor Constantine, researched this same area southeast of Bethlehem in around 325AD. She also came to similar conclusions as I did regarding the location of the Shepherd's Field and was so convinced she founded a monastery on that very site. She named it the "Meadow of the Shepherds." Today, there are three sites vying for the official title of the Shepherd's Field, all are within a short distance of each other.

Location of the Shepherd's Field East of Bethlehem

Where, Oh Where did the Star of Bethlehem Go?

It all seems to fit. Either this is the hugest coincidence in history or this may actually have been what happened. In the early hours of the morning of January 8, 4 BCE, Jupiter was setting in the west in conjunction with the Moon and the progenitor star of M1. All three celestial objects were in a straight line as anticipated and these Astronomers were in bliss. But something was about to happen that even these scientific geniuses could not have predicted. As the first hints of morning approached, the M1 progenitor star flared even more brightly and from their vantage point Bethlehem was directly below.

The star Biblically recorded as the Star of Bethlehem was in fact the progenitor star of M1 in its beginning death throws. It was either flaring up or changing color and announcing its impending demise that would inevitably occur over 1,000 years later. I suspect a color change may have been the first sign observed back in Persia and a significant change in brightness ("momentary" flare) noticed in Bethlehem.

Some scientists estimate that this momentary flare may have been as much as10 times brighter than the planet Venus. This would make it by far the brightest thing in the sky that night. I believe that these Magi, being curious men of science, looked for other clues to understand more about why this particular star flared above this particular city at this particular time.

But the classic story of the Star of Bethlehem has one more twist as we shall see in the next chapter.

Exit...Stage Right...

As the Magi watched the conjunction from the Shepherds Field in the wee hours of the morning, the progenitor star of M1 flared to an intense brightness just before setting on the western horizon. I can only imagine what the dialog was like on that hilltop field between the two groups of people separated by language but not by common experience. The excitement must have been contagious where both camps (Herod's spies and the Magian's) could barely believe their eyes. I can imagine them pointing up, using frantic language as they called to each other, and everyone standing tall with that small surge of adrenaline that makes such frenzied experiences unforgettable. But the Biblical account goes just a little further in disclosing one last thing about that magical moment: a visit with a child.

The book of Matthew makes no mention of anything else occurring that night unlike the account in the book of Luke where Angels appear announcing the birth of Jesus to the Shepherds. The account in Luke is the one that has traditionally been accepted as what transpired but chronologically speaking the book of Matthew was written prior to Luke. Setting these differences aside and focusing only on what is recorded in Matthew, let's put one more scenario together that seems to make sense.

Let's regroup for a moment and think about what just happened from the perspective of these non-Hebrew scientists. The event ended before dawn and it was time for the Magi to end the expedition and go back home. The whole trip was a huge success and the Magi could not wait to tell King Phraates IV about their findings. Their instruments survived the journey and they made all of the measurements they desired – and then some! After back slapping, high fives, and the like they eventually settled down. Their attention must have turned to packing up, getting some sleep, and planning the return trip back to Aqaba. The mood in the Magi camp must have been one of joy and celebration. If they had not come, they would not have observed the last brightening of the progenitor just prior to its setting. This observation alone would thrill King Phraates IV knowing that he had backed a winning horse – so to speak. The mission was over; the Magi could relax; and everyone could get some much needed rest.

The next day, the Magi would have taken down their instruments, loaded up their pack animals, and prepared to leave. Whatever spies King Herod employed to monitor their activities had no reason to stay so they too packed up and went home. While others were packing up, some of the expedition's support staff probably wandered into Bethlehem to pick up whatever last-minute supplies were needed. As it is in any conversation with strangers, these local merchants most likely inquired about the return route they planned to take. Being locals, these merchants knew the "short cuts" and their conversation switched to that of an alternate route. Rather than going back the way they came through Jerusalem, they could cut off several days of travel by taking a different trail. I can easily envision the thought of getting home sooner as being an attractive option to everyone. Before we continue down this non-traditional avenue of speculation, it would be a great time to recap everything.

Astronomers from Persia saw something in the night sky that was so startling they convinced their King to fund an expedition to the Mediterranean Sea. This King was convinced that this event was unique enough to warrant his investment and agreed. Many adventures were encountered in getting to their destination, one of which was encountering a 2,500-foot tall range of mountains that blocked their route to the sea. Undeterred, the expedition joined a caravan headed for Jerusalem. Upon their arrival, King Herod wanted to know if these foreigners were spies. After meeting with them and hearing their story, he interpreted what they said as a personal threat to his ongoing authority. After another meeting with his people (some of which were spiritual leaders), he found that there was a prophecy about a child-king to be born in Bethlehem. King Herod decided to allow travel for this expedition to Bethlehem so that in all probability they would find this child. The Magi arrived in Bethlehem and set up their instruments in the best location they could find: a field east of town. As they observed the event, M1's progenitor star grew in unusual brightness as they watched it sink below the horizon appearing as if it touched the rooftops in town of Bethlehem.

But there is one more adventure ahead of this expedition, one more side jaunt that also made it into the classic tale.

It is Biblically recorded in Matthew 2:12 that the Magi "…left for their own country by another road…" This implies that there is another route across the 2,500-foot tall mountains between Bethlehem and the Dead Sea, however, today there exists no such road. But searching for ancient maps of the region on the internet, I discovered one showing this "short cut" across the mountains that appeared to be fairly direct.

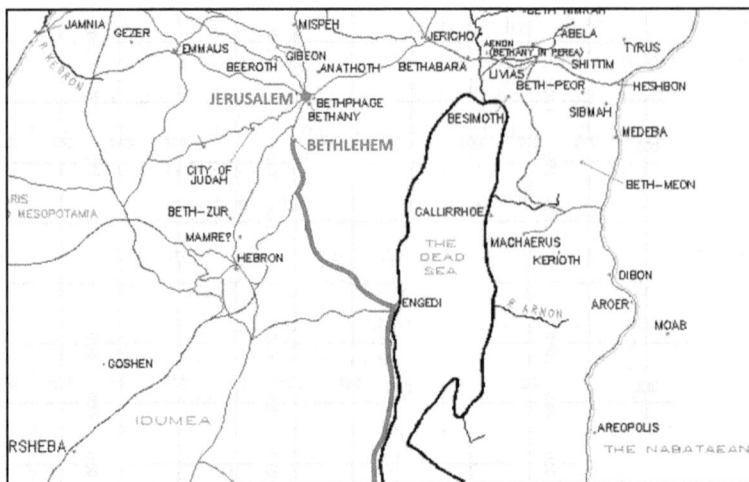

Magi Return Route via Engedi (Ein Gedi)

It appears that the possibility of a local telling the expedition of a short-cut may very well have happened. But let's make sure that such a route exists and could even be accomplished.

To verify this route, I used the satellite feature of Google Maps™ and so can you. Just north of Ein Gedi, there is a riverbed called the Nahal Arugot that winds its way westward up to a summit. At some point along the way, this riverbed crosses an existing rural road, 3698. Following 3698 westwards it connects with other existing roads and eventually leads you to Bethlehem via road 3686. Taking an expedition along a precarious stretch of riverbed would be a rough route to travel but it would be passable albeit slow going. Since there were no time constraints for returning home, it wouldn't matter how long it took for them to cross this stretch of mountains, especially if it meant getting home much sooner.

The remnants of this alternate route may be the same route the Magi used instead of going back through Jerusalem. It seems logical and, from the verified route, even plausible. So how did they find the child as described in Matthew? Here is what I suspect happened.

The expedition climbed down from the Shepherd's Field on the same trail they used to reach its height to restock supplies. This trail led back toward Bethlehem but the short-cut route they were to take led away to the southeast. Along this alternate route, the trail would take them into the foothills and then into the mountains. There are caves in the area along this alternate route not too far from Bethlehem itself.

Where, Oh Where did the Star of Bethlehem Go?

What if somewhere along their alternate route they accidentally ran into a cave where a young family was staying with a newborn child? Remember, their original astrological interpretation of this celestial event marked the birth of a new King. Would the Magi have connected the dots? Now <u>that</u> would be a truly amazing coincidence!

Perhaps the entire trip occurred as I have presented; perhaps not. Once you pay attention to human nature and inject typical behavior into a well-accepted belief, especially one that has seriously suffered from commercialization, is it possible that it all went down this way or with some minor deviations to include other accounts of this same tale? Regardless of how you react or what your beliefs are, you must admit if nothing else it is an interesting scenario with some astounding coincidences. What do you think?

What Did the Original Words Say?

Before we jump to conclusions about something, let's try to be pragmatic about the event approaching it again from a detective's perspective. A professional desires to separate the facts from a case by removing any emotion, suspicion, innuendo, and hearsay. Witness statements taken at the scene of a crime will always vary depending upon many things from viewing angle, attention to detail, or preoccupation to name just a few.

And so it is much the same with translations. Each individual performing the translation has certain biases or opinions that influence word choice and interpretation based on personal knowledge and their intent. This is the natural order of things and it is why legal cases are sometimes so complicated to understand and evidence so difficult to interpret.

The great thing about any book written in a foreign language is that there are so many translations from which to choose. Each person, when making that translation, seriously tries to understand not only the literal meanings of the words but also any local deviations and slangs associated with them in use at the time of the writing. This is a daunting task and this is why there are so many versions of the same thing.

Of course it is best to use the original version when doing any translation or interpretation. The oldest *surviving* copy of the work of Matthew is written in Greek, but there is some serious doubt that this is truly the original version. Many scholars, far more learned than I on this subject, believe that a smaller version was written prior to the Greek in the Aramaic language. However, there are no surviving copies and we must unfortunately use what we have.

Two people, Brooke Foss Wescott and Fenton John Anthony Hort, preserved the surviving Greek words of the Bible's New Testament in their 1881 *Harper and Brothers* publication entitled "The New Testament in Greek." It is also known that Wescott did not believe in miracles; I respect that since he would most likely not bring a spiritual bias into their translation but rather an objective one.

I personally believe it is important for you to decide on your own rather than taking the words of the so-called experts (we've already seen how so-called experts can be proven wrong). So I have included excerpts from the Wescott and Hort book to allow you to decide for yourself what the words meant. I encourage you to use other references in an attempt to glean the true meaning. And remember, even though you may understand the content today, you may not understand the context of the language itself (those unique idioms and nuances similar to that of today's slang).

Where, Oh Where did the Star of Bethlehem Go?

Original Greek	Westcott and Hort literal translation
Matthew 2:1	
tou de ieesou genneethentos en beethlEEm	of the but Jesus having been generated in Bethlehem
tees ioudaias en heemerais heerwdou tou basilews idou	of the Judea in days of Herod the King. Look!
magoi apo anatolwn paregenonto eis	Magi from eastern parts came to be alongside into
ierosoluma	Jerusalem
Matthew 2:2	
legontes pou estin ho techtheis basileus twn ioudaiwn	saying where is the (one) born king of the Jews?
eidomen gar autou ton astera en tee anatolee kai	We saw for of him the star in the east and
eelthomen proskuneesai autw	we came to do obeisance to him.
Matthew 2:3	
akousas de ho basileus heerwdees etarachthee kai	Having heard but the King Herod was agitated and
pasa ierosoluma met autou	all Jerusalem with him,
Matthew 2:4	
kai sunagagwn pantas tous archiereis kai	and having led together all the chief priests and
grammateis tou laou epunthaneto par autwn pou	scribes of the people he was inquiring beside them where
ho christos gennatai	the Christ is generated.
Matthew 2:5	
hoi de eipan autw en beethlEEm tees ioudaias	the (ones) but said to him in Bethlehem of the Judea;
houtws gar gegraptai dia tou propheetou	thus for it has been written through the prophet
Matthew 2:6	
kai su beethlEEm gee iouda oudamws elachistee	and you, Bethlehem land of Judah, by no means least
ei en tois heegemosin iouda ek sou gar	are in the governors of Judah; out of you for
exeleusetai heegoumenos hostis poimanei ton	will come forth governing one, who will shepherd the
laon mou ton israeel	people of me the Israel.
Matthew 2:7	
tote heerwdees lathra kalesas tous magous	Then Herod secretly having called the Magi
eekribwsen par autwn ton chronon tou	carefully ascertained beside them the time of the
phainomenou asteros	appearing star,

Original Greek	Westcott and Hort literal translation
Matthew 2:8	
kai pempsas autous eis beethlEEm eipen	and having sent them into Bethlehem he said
poreuthentes exetasate akribws peri tou	having gone on way search you carefully about the
paidiou epan de heureete apaggeilate moi	young child; whenever but you might find report back to me,
hopws kagw elthwn proskuneesw autw	So that also I having come might do obeisance to it.
Matthew 2:9	
hoi de akousantes tou basilews eporeutheesan	The (ones) but having heard of the king went their way,
kai idou ho asteer hon eidon en tee anatolee	and look! The star which they saw in the east
proeegen autous hews elthwn estathee epanw	went ahead of them, until having come it stood above
hou een to paidion	where was the young child.
Matthew 2:10	
idontes de ton astera echareesan charan megaleen	Having seen but the star they rejoiced joy great
Sphodra	very much.

Fascinating! As you can see, there is a lot of room for interpretation. By examining the literal translation of Wescott and Hort, you can glimpse the difficulty in trying to properly convey the meaning of a different language from a different time; it is challenging to say the least.

What Would You Carry?

One big issue around leading an expedition of this magnitude and duration is timing: when do you leave? To assure that the members of this expedition would reach their destination in time to observe what it is they wanted to see, they would have to allow enough time to get there. Traveling lighter lends itself to traveling faster. Even though they would have preferred to take all of their instruments, toting large or heavy items all that way would just not be practical, especially if timing were a concern.

Taking a lot of instruments means a larger and more costly expedition so their selection would involve not only what they needed but also how that would impact their budget. And then there is the issue of arriving at their destination, having enough time to set up everything, and still have plenty of time to thoroughly observe and measure the event. One thing is for certain: they did not want to rush through the observation of this event and make a silly mistake.

A few of the more portable astronomical instruments that may have been used at this time include the Antikythera Mechanism developed in about 150 BCE, Gnomon (Sundial), Armillary Sphere, Astrolabe, and sextant-styled tripod-mounted device used for angular measurements called a *Dioptra.*

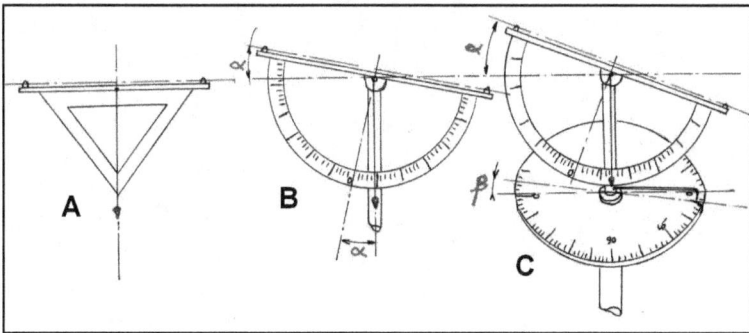

Wikimedia Commons Image of early Dioptra by Nerijp (Own work)

Of course the Magi would also need their reference materials, charts, drawing tools, inks, papers, tables, chairs, lamps, tents, and similar supplies to properly document what they observed. If you have ever backpacked into the wilderness, you know that each ounce you carry slows you down so everything they took with them mattered. If a pack animal died on the trail or succumbed to an injury, whatever it was carrying would have to be left behind; spare animals added to the cost of the expedition and may have been included if the budget would bear it.

Having backpacked in both summer and winter in my younger years, my personal opinion is that they traveled <u>very</u> light, took the barest necessities, and planned ahead for misfortune. What do you think? How big of an entourage would your expedition be? How much extra food, clothing, footwear, and water would you carry? Where would you stay when it rained or snowed? How much gold would you need (the heaviest item you would carry by far)? Answering these simple questions will give you a rough understanding as to how big this expedition truly was.

Creating your own list of equipment and planning your own expedition will give you a deeper insight as to what these brave travelers endured. It will help you appreciate how much effort, persistence, and perseverance was required to commit to such an incredible adventure. I hope you do this little exercise just for you own personal insight because trying to relate to their situation helps you understand how the translations and oral traditions may have altered the recorded Biblical text.

Reviewing the Facts

Other than the account in the Bible of the Star of Bethlehem, there is no mention of it in any other secular records or even in subsequent books in the Bible; this Star just disappeared. Many theories have tried to tie this Biblical account to scientific fact and all have fallen short, until now. Let's review the facts.

Chinese astronomers documented the demise of the progenitor star of M1 as occurring in 1054 creating the well-known Crab Nebula. In what is called a supernova, a star ends its life in a spectacular fashion. One day, the star looks like any other nearby star and the next day it is a bright beacon.

In 1987, astronomers at the Australian Astronomical Observatory (AAO) photographically recorded what a different star (SN1987a not M1) looked like before and after it explodes (see http://203.15.109.22/images/captions/aat049.html).

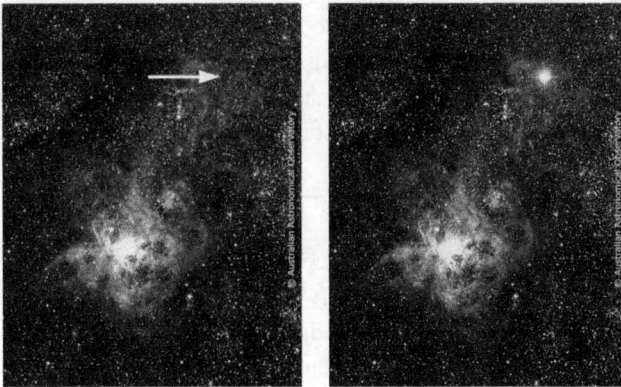

A Supernova One Day Before and After Exploding
Photographs by David Malin

Astronomers identified the progenitor star of SN1987a as *Sanduleak -69° 202*. It encountered a Type II "Peculiar" supernova, atypical for the behavior of a blue supergiant star. The word *Peculiar* means in some way it did <u>not</u> follow the expected behaviors of an otherwise 'normal" Type II supernovae whose remnants create a neutron star. SN1987a did not create a remnant neutron star, hence this "Peculiar" category.

Astronomers also know that M1's progenitor star encountered a Type II supernova meaning its behavior was a little different from that of SN1987a. The M1 supernova created an optical pulsar, a visible rotating remnant neutron star spinning at over 32 times per second.

All Type II supernovae change color and/or brighten prior to exploding. Studies of the current mass of the remaining Crab Nebula against the projected mass of M1's progenitor star do not add up. This means there is missing mass and indicates that something else probably happened to this progenitor star before its final explosion in 1054 (see http://www.sponli.com/en/Object/Index/2990).

One possible explanation for this missing mass is that M1 experienced <u>two changes</u> in color and/or brightening, one in 1054 and another some time earlier. I suspect this was the case where a change in color occurred in 7 BCE and observed by the Magi from their home observatory in Persia; a change in brightness occurred in 4 BCE and where the Magi were waiting for a triple conjunction on a hillside southwest of Bethlehem; and its final demise was observed by Chinese astronomers in 1054. The progenitor star of M1 was in the right place at the right time and its color and brightening behavior is what I suspect the Magi observed.

The truth about where M1's missing mass went is documented as unknown. However with the discovery of multiple changes in color and/or brightening of a progenitor star, it could explain this missing mass. Two changes in color and/or brightening of M1 is a plausible explanation of where this mass went (mass was converted into light energy during its first change).

What Does This All Mean?

The probability of a Biblical reference being reasonably historically accurate has been viewed by many academics as *highly unlikely*, however in this case it appears to have credence when making certain adjustments attributed to human error and misunderstanding. By the documented admission of the Catholic Church, the current calendar year is incorrect by 1-7 years (here the midpoint of four being the operative number). Our understanding of supernovae behavior changed; initially believed only one brightening occurs is now proven to be incorrect. Understanding that the common view of the Magi as being astrologers is also incorrect (they were scientists and astronomers). Asking the right questions about this event changes the perspective about stellar candidates (why did the Magi have to travel so far to see this event and not just stay home?).

Collectively, these data imply each other to be likely where taken individually there is no correlation at all. In other words, after putting all of the pieces of this puzzle together, M1's progenitor star as a candidate for the Star of Bethlehem makes perfect sense.

This entire analysis has an interesting parallel to a childhood game. Take a row of ten people, whisper a short phrase into the first person's ear and then have the each whisper that phrase to the next. What happens is that the phrase heard by the first person never exactly matches the phrase heard by the tenth person. Such is human behavior and regardless of how careful we are human beings will always make mistakes. Even if one version of an ancient language is perfectly preserved, our contemporary interpretation of the words changes (think of the changed meaning of the word "gay" as primarily understood in 1920 and in 2015; once it primarily meant *happy* and now it primarily implies *homosexuality*). Scholars interpreting ancient texts have similar issues when agreeing on how old texts should be translated.

As a writer, I understand how poetic license creeps into content; I do it all the time and I bet you probably do to. For example, almost everyone does this same "editing" when talking to children (or to someone who does not understand the subject). When trying to describe a complicated thing such as an adult issue to a child, you will likely make a translation that resembles the original but uses far less-complicated words (exactly the same thing only way different!). Even when passing on a joke, rarely is it told exactly the same way.

What remains in a translated version is the "spirit" of that description; this is what I searched for in the sky when trying to identify what this Star of Bethlehem could be. For example, did the words translated as "go before them" mean the same thing thousands of years ago as they do today? I doubt it. But does this mean that the "spirit" of these words is wrong? Probably not. Understanding the hidden message in any tale is the trick to understanding the original meaning of their words; including the context helps you determine what they may have really meant. Think for a moment about someone being misquoted. Taking things out of context can change everything so context is very important in truly understanding.

Is the Messier Object designated as M1 the remnant of the Star of Bethlehem? I believe it is. Were the Magi astronomers and students of science? I believe they were. Was Jesus born on or before January 8, 4BCE? If the context of the tale of the Star of Bethlehem has survived 2,000 years of poetic license, then yes, I believe he was. Does it make a difference to you? Perhaps, but it may also reinforce what you currently believe to be true. Wouldn't it be great to able to see the same stars today that the Magi did so long ago? The good news is that M1 is still visible today and the next chapter explains how you can see it.

Where Is This Star Today?

The best time to look at the stars is when the moon has gone down or it has yet to rise (near the "new moon"). During this time the sky is its darkest and you can see more of them (the light of the moon does not wash out the night sky). The best time to see M1 in December is the weekend of December 24.

At about 8:00 PM local time on the early evening of December 24, 2016, M1 will appear above the horizon almost due east. If you go out later in the evening, it will appear higher in the sky. M1 is to the north of (to your left of) the constellation Orion, an easily-spotted star pattern in the sky. Orion has three bright stars in a row that make up his belt; others make up his shoulders, head, and feet. The distance from Orion's top star to his belt is about the same distance from Orion's top star to M1.

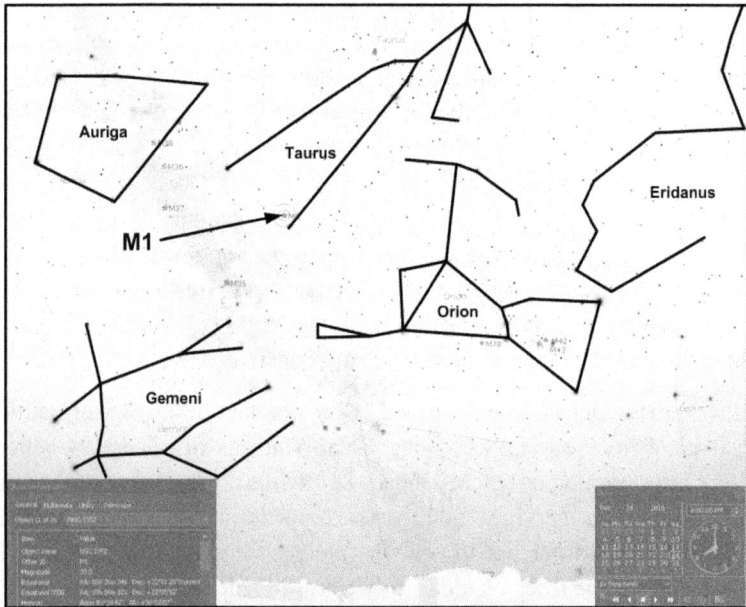

The Location of M1 on December 24, 2016, at 2000 Hours
Looking Due East from Tampa, Florida

M1 is a faint smudge-like object you can see under dark skies with a modest sized pair of binoculars. Through a 10x50 binocular, the view would look something like this:

M1 as Seen through Small Binoculars

If you slowly scan the sky on any moonless night above the top star of Orion, you should be able to find M1. The next time you spot it, think about this nebula as being in the same place in the sky that the Magi looked over 2,000 years ago.

And on April 1 of 2017, the Moon will be near M1 (in conjunction with) in the western sky at about 8:00PM local time. Although not identical, it will be similar to the conjunction the Magi saw so long ago.

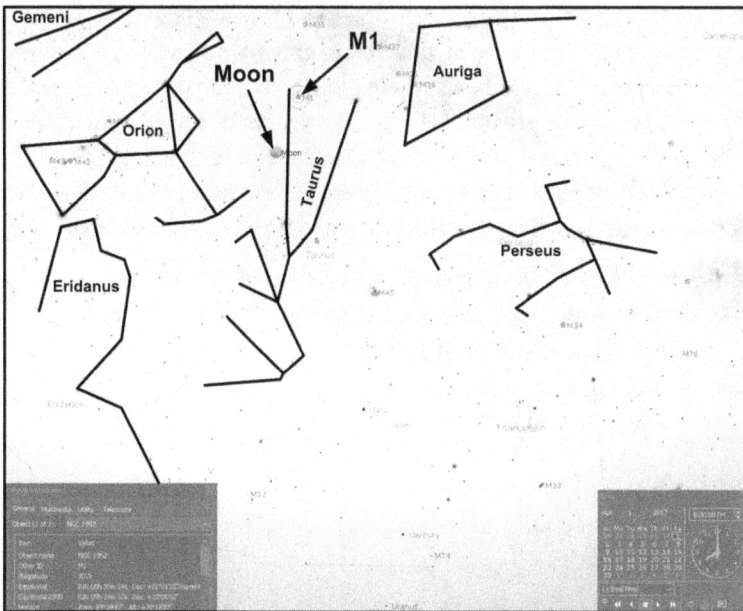

M1 and Moon conjunction on April 1, 2017, at 2000 Hours
Looking Due West from Tampa, Florida

I want to thank you for buying this book and bearing with me while I explored the depths of human nature and scientific facts. I hope that you found some points of interest and I do not expect that you would believe everything you read, especially when it comes to your personal spirituality. Like a smorgasbord, take what you like and leave the rest. This is one possible explanation, one with many coincidences, and an explanation that I hope you enjoyed.

Whatever you choose to believe is just that, your choice. However, remember that what you believe as truth today may not be true tomorrow. Change is inevitable as is our understanding of how our universe works. The doctrine of the Roman Catholic Church once believed that the world was flat and located at the center of the universe and those who disagreed with these doctrinal beliefs were persecuted or even put to death. Regardless of what they spiritually believed, these ancient spiritual leaders were wrong even though their beliefs firmly convinced them that they were right. Change eventually came to these core spiritual beliefs as well.

Is it important for you to believe that the Star of Bethlehem existed? Who knows? Will our scientific understanding of this event change? Definitely! Are my assertions correct? Perhaps; perhaps not. But it is interesting to note the huge number of coincidences I have uncovered and how they re-enforce each other.

In understanding what happened so long ago with no clear supporting evidence to confirm much of anything, this is where logic exits and your spiritual beliefs enter. For a moment, drop your existing beliefs and consider this hypothetical adventure. It is fascinating to speculate what may have happened so long ago and why people are still talking about it today. I encourage you, if nothing else, to talk to your own spiritual leaders and understand exactly what and why you believe what you do.

As for my fellow astronomers out there, what do you think? I know my conclusions would collapse under detailed scrutiny – or in a court of law. But is it *possible* that a Biblical, non-scientific reference to a "star" is reasonably accurate when considering my assertions? When making minor adjustments and concessions, do the puzzle pieces fit? Something to think about...

About the Author

Philip Rastocny developed personal spirituality from a very early age. Raised in a religiously-grounded ethnic family and attending parochial schools through high school, he was trained daily by learned instructors about their interpretation of Biblical texts. Fascinated by the reality presented by his senses, he explored the mysteries of life and searched for the answers to classic questions that his family's religion did not explain to his satisfaction like *Why am I here?* and *What is life all about?*

In his senior year in high school, he took a class in astronomy and became hooked on the wonders of the night sky. Laying on the soft grass in the backyard his rural home in Wisconsin, the warm summer evenings and clear skies provided endless hours of entertainment. His father's binoculars gave him a glimpse into a reality filled with wonder and awe, a reality that was always there waiting patiently for him to discover.

After serving in the United States Air Force and graduating from Oklahoma State University, he and his loving wife moved to Colorado and built a home high atop a 9,500-foot mountain. There he trained his many telescopes into the cold deep darkness overhead. But one question gnawed at him without answer: *What was the Star of Bethlehem?* Decades of research later, answers to seemingly unrelated questions began to "add up." One day, when he least expected it, a PBS special about the Silk Road rekindled his need to search again...search for an answer to that unexplained star.

Approaching the problem like a detective solving a murder mystery and considering human nature in their motives and justifications, he slowly fit each puzzle piece together. With the purchase of an astronomical computer program that allowed him to recreate the night sky from any place and any time in history, he tested his theories and developed scenarios that could explain many of the questions he had.

Although finding a deeper understanding of the events in the story of the birth of Jesus, it wasn't until an accidental discovery while scanning the heavens from his own theories that everything fell into place. All of his previous knowledge in the other areas of his life came together simultaneously to provide an insight that no one else had considered. Like discovering who committed a murder in a complicated novel and how it was achieved, each piece locked together in an almost melodic composition. The light turned on!

Philip had always read the account of the Magi in the book of Matthew from a spiritual perspective with the intent of reinforcing his childhood beliefs. But after applying human nature and distancing himself from the religious biases of his instructors, he could see more clearly what the brief description meant and what actually transpired. Like reading the Cliff's Notes version of a book, he never really thought about what each word in the book of Matthew meant – until now.

It did not strengthen or weaken his spirituality to uncover what the Star of Bethlehem was or where it went, it only answered a few of those gnawing intellectual questions and gave him peace. The next time you look up at the night sky think about this mysterious star, these ideas of what it was, and what that means to your own personal spirituality. Philip hopes that this story gives you the same peace he now has.

www.ingramcontent.com/pod-product-compliance
Lightning Source LLC
La Vergne TN
LVHW051153080426
835508LV00021B/2608